# Archaeological Remote Sensing in North America

# Archaeological Remote Sensing in North America

## Innovative Techniques for Anthropological Applications

Edited by
Duncan P. McKinnon and Bryan S. Haley

Foreword by Jay K. Johnson

Commentary by Kenneth L. Kvamme

The University of Alabama Press
*Tuscaloosa*

The University of Alabama Press
Tuscaloosa, Alabama 35487-0380
uapress.ua.edu



Typeface: Caslon and Gil Sans

Manufactured in the United States of America
Cover image: Ancestral Mandan lodges at Huff Village revealed by magnetic gradiometry and elevation data; courtesy of Adam S. Wiewel
Cover design: Erin Bradley Dangar / Dangar Design

Cataloging-in-Publication data is available from the Library of Congress.
ISBN: 978-0-8173-1959-5
E-ISBN: 978-0-8173-9141-6

# Contents

# Illustrations

# Foreword

Since the 2006 publication of *Remote Sensing in Archaeology*, which I edited, there has been a major sea change in North American archaeology. Most major academic programs in the Southeast and several cultural resource management (CRM) firms have acquired at least one geophysical survey instrument, usually a gradiometer but sometimes ground-penetrating radar (GPR). This volume is an elegant testimony to the ways that remote sensing has changed how we do archaeology.

Of the bold predictions about the direction of the field that I made in *Remote Sensing in Archaeology*, the one that was most likely to come to pass was that the technology would improve. For example, from 2000 to 2010, Bryan Haley and I relied on a 30-foot-long helium-filled blimp for low-level imagery or else borrowed Tommy Hailey and his powered parasail. Now anyone with a credit card can collect images of the White House lawn. While the thermal camera that we used was too heavy for most commercial drones and was about the size of a 1990s camcorder, current thermal imaging systems have become much smaller and cheaper. Agema, the company that made our sensor, is now advertising in sailing magazines compact systems that plug into an iPhone like a credit card scanner. Also, several next-generation geophysical survey instruments are now available that make it much easier to collect data. Many of these instruments were built in response to some of the challenging rituals that one had to perform to use the older instruments. Anyone who has stood on a nonmagnetic step stool in the middle of an open field, turning to face the cardinal directions in seemingly endless rotations, will surely appreciate the ease with which the Bartington Grad601 gradiometer can be zeroed. There is, of course, no comparison between the notebook computers of 2004 and those of today in terms of memory, storage, and computing speed. The case studies that follow all benefited from these advances.

There has been a subtle shift in the relative popularity of the various geophysical techniques that are being used. For reasons that are obvious from some of the spectacular images of large, complex Mississippian period mound sites or equally complex Plains Village sites, gradiometry has become the go-to method. GPR continues to be important and, once again, as several of the following case studies demonstrate, has been used with increasing success in producing three-dimensional models of mound structure and other archaeological deposits. While much more demanding in terms of instrumentation, application, and data processing, electrical resistivity tomography has yielded some remarkable results and is particularly suited to exploring mound structure, providing a welcome adjunct to GPR.

This brings me to one of my predictions that failed to pan out. On the basis of our encouraging results in measuring magnetic susceptibility using an EM38B, as well as Rinita Dalan's pioneering work in developing the Bartington down-hole sensor, I expected magnetic susceptibility to play a larger role in the growth of geophysical remote sensing. Although a core hole must be made to use the down-hole sensor, it provides a relatively inexpensive way to recover three-dimensional data. I expected the EM38B to be a much more popular instrument because it made it possible to gather both conductivity and susceptibility readings simultaneously. As Ken Kvamme points out in the commentary to this volume in part 5, horizontal mapping of susceptibility is valuable in that it allows us to isolate one of the two aspects of magnetism that are undifferentiated by the gradiometer: remanant magnetism and magnetic susceptibility. Each is the result of different kinds of human activity and is likely to detect different kinds of features. The only way to distinguish between, say, pits filled with organic-rich soils and pits filled with daub using a gradiometer alone is to look at the strength of the signal.

Another trend that I anticipated that has not fully developed is multimethod surveys and data fusion. Ken Kvamme contributed four of the five case studies to the chapter dealing with this topic in the 2006 volume and has been a frequent and persuasive advocate for using several different instruments when doing a geophysical survey of an archaeological site. Several of the chapters that follow report on the use of two or more instruments. However, the identification of different kinds of features in the imagery is almost exclusively dependent on spatial characteristics. Several contributors call for the need to develop signatures that will allow us to go beyond shape and size, but there is little effort to take advantage of the possibilities inherent in applying one of the many multivariate statistics that are routinely used in dealing with multispectral satellite and airborne imagery.

Before a NASA grant in 2000 that allowed me to expand my interest in remote sensing to include several geophysical survey instruments, my major research emphases were lithic analysis, the Colonial Chickasaw, and the

Archaic period. Thereafter, I shifted my research interest to the Yazoo Basin of the Mississippi River floodplain. The results that my students and I have obtained using a gradiometer on Mississippian period sites such as Hollywood, Parchman, and Carson provided rich data for 15 years of conference presentations, journal articles, and theses. We have spent much less time on the Archaic and Woodland period sites in the same region, in part because they do not contain wattle and daub houses that, when burned, are easy to detect using a gradiometer. I believe that we will not be able to successfully apply geophysical survey techniques to these kinds of sites unless we use two or more instruments and process the data using multivariate statistics.

In the commentary at the end of this volume, Ken Kvamme discusses the relative popularity of geophysical surveying in the Southeast. I would like to think that the 2002 NASA-sponsored workshop in Biloxi that Marco Giardino and I hosted (before the annual Southeastern Archaeology Conference meetings) had a role in this, but I expect there are a couple of other factors. There were three relatively early university-based programs that integrated geophysical survey techniques and archaeology in the Southeast, with Erv Garrison at Georgia, Ken Kvamme at Arkansas, and me at Mississippi. Berle Clay at Cultural Resource Analysts, Inc., also played a role in the propagation of the techniques in the Southeast. He introduced me to conductivity surveying at the Hollywood site in 1997 and came back with a gradiometer in 1998. During lunch on his first day at the site on his second visit, he processed four 20 m grids of gradiometer data that centered on a house floor that we had found and tested using his conductivity data. Relatively few keystrokes in the shade of the trees at the edge of Mound A produced an image that clearly delineated the wall trenches that we knew were there. I decided at that moment that we had to have a gradiometer. Finally, while satellite and airborne sensors are not nearly as useful in the heavily vegetated Southeast as they are in the Southwest, geophysical surveys at sites such as Kincaid, Moundville, and Poverty Point have produced complex images that are essentially site maps, allowing a great deal to be learned by digging precisely targeted "ground-truth" test pits. In fact, as this present volume demonstrates, the quality of the data that can be derived from geophysical surveys makes it possible in many cases to bypass excavation altogether and go directly to analysis.

I close by addressing the question of whether the goal that motivated the 2002 workshop and subsequent book has been reached. That is, has remote sensing become an integral part of CRM archaeology in North America? One measure is the contract work done by the Center for Archaeological Research at the University of Mississippi. Starting in 2001, I have had the pleasure of working with a number of research associates, all of them recent MA graduates from our program: Bryan Haley, followed by Ed Henry,

Stephen Harris, and most recently Travis Cureton. They were all supported by contract work. Of the 56 contracts we worked on during that period, nearly all included geophysical remote sensing to some degree. Nearly half of those, 26, were CRM contracts but—and this is telling—more than a third of the CRM contracts, 9, dealt with historical cemeteries that were threatened by road construction. We did far more work on archaeological sites that were not cemeteries outside the CRM context, and this does not count the 15 years of field school that always included geophysical surveys. For much of this period we were the only research entity in Mississippi that did remote sensing on a regular basis. If remote sensing had become an integral part of CRM archaeology, we surely would have had more opportunities to do CRM-based geophysical surveys.

Of course, this book provides a different opportunity to assess the impact of remote sensing on CRM archaeology. Eight of the thirteen following case studies report on CRM-based archaeology. Yet it is an entirely different kind of CRM from what we had in mind when we put together the workshop in 2002 and subsequent book. In none of these cases are the cultural resources being investigated threatened with destruction by a federally regulated construction project. Rather, the cultural resources are located within state or federal parks and the geophysical surveys were done to enhance our understanding of those properties. Most of the contributors discuss the obvious value of remote sensing research in dealing with protected cultural resources. That is, we maximize the amount of information that can be recovered while minimizing the impact on the sites. One contributor, Philip B. Mink II, in chapter 4, goes so far as to propose that whenever the National Park Service permits excavation on its property, it should require remote sensing. I like that idea a lot.

Although this is not the future that I predicted, as this book demonstrates, it is a good place to be. Not only is there significantly more remote sensing being done, but the kinds of questions being asked and answered are a lot more interesting. Geophysical remote sensing continues to be an important tool in developing a research design, allowing excavation to be located with the kind of precision that surgeons derive from studying X-rays, magnetic resonance imaging, and other imagery. This anticipates a point that Ken Kvamme details in the final chapter, "A Decade of Geophysics and Remote Sensing in North American Archaeology": we are now doing so much more with geophysical survey results. In many of the chapters that follow, geophysical imagery is not used to supplement other kinds of data; it is the focus of analyses that explore a wide range of anthropological issues.

Jay K. Johnson

# Preface

*Duncan P. McKinnon and Bryan S. Haley*

This volume began in the same way that Jay Johnson's 2006 edited volume *Remote Sensing in Archaeology* did, over a cold beer. In this case, we were recovering after a long and arduous day of geophysical data collection as part of the PBS television show *Time Team America* in the summer of 2012. After a number of conversations about our individual experiences, the proliferation of archaeological remote sensing, and the application of several exciting new trends, we thought that the timing was right to put together a new edited volume, this time using a set of case studies to illustrate new developments in archaeological remote sensing and its role in anthropological applications. The very fact that these discussions happened over the course of filming *Time Team America* is noteworthy and is a sign of the changes that had taken place in North American archaeological remote sensing in the years since the publication of *Remote Sensing in Archaeology.*

The roads that led each of us to be involved with the present edited volume are worth reviewing. I (Haley) was an MA student under Jay Johnson at the University of Mississippi. With a background in computer science, math, and physics as an undergraduate at the University of Kentucky before ultimately deciding to study anthropology, I was lucky enough to hear about the initiation of a geoinformatics fellowship program at the University of Mississippi. For their MA theses, geoinformatics archaeology students would be trained in Geographic Information Systems (GIS), airborne remote sensing, and geophysical techniques, while having access to data from the earth sciences wing of NASA's Stennis Space Center, directed by archaeologist Marco Giardino at the time. The other geoinformatics students and I spent the next two years learning various geospatial technologies, in addition to our normal anthropology regimen. During our thesis research, which was conducted at the Hollywood site (see chapter 3) in the summer of 2001, we were instructed by some of the leading practitioners of archaeological remote sensing, including Bruce Bevan and Ken

Kvamme, as well as Giardino. Once I finished my MA in 2002 (see Haley 2002), I remained at the University of Mississippi to help Jay Johnson with the Center for Archaeological Research, where I assisted in teaching students remote sensing techniques and conducting surveys across the Southeast and beyond. As a result of that experience, Johnson invited me to contribute to three chapters in the 2006 publication. Another highlight was being invited by Meg Watters, a colleague and my first GPR instructor, to be a part of *Time Team America*, the first season of which was filmed in 2008. It was during the second season that I met and worked with McKinnon.

After spending 10 years working in high-tech hardware and software, I (McKinnon) decided to pursue my interest in history and began undergraduate studies in anthropology at Texas State University under the guidance of Kent Reilly. During my first summer break, I and a few other students were invited to help conduct geophysical fieldwork at the Etowah site under the direction of Kent, Adam King, and Chet Walker. This was my first introduction to archaeological geophysics. The results were spectacular (see A. King et al. 2011; Walker 2009), and I was hooked. I spent the next seven years at the University of Arkansas under the tutelage of Ken Kvamme and others from the Arkansas Archeological Survey and the Center for Advanced Spatial Technologies while completing my MA and PhD. During my tenure at the University of Arkansas, I was trained in and exposed to numerous geophysical techniques and GIS applications in archaeology while working on projects throughout North America (see Kvamme 2008a; Kvamme et al. 2009; Kvamme and Wiewel 2010; McKinnon and Kvamme 2007; Wiewel et al. 2009), including the Battle Mound site, where my doctoral research was conducted (see chapter 3; McKinnon 2013a; McKinnon 2017). One afternoon, I received an e-mail from Kvamme about a second season of *Time Team America*, which was to be filmed the following year. His former MA student Meg Watters (Boston University) was looking for an additional geophysical team member for the second season. After a lengthy phone call with Meg, I was invited to be a part of *Time Team America*, where I met and worked with Bryan Haley and we began the discussions that ultimately led to this volume.

Over the course of our training and application of geophysical methods during the last several years, we have had the privilege of working and interacting with many of the authors included in this book. These professional and personal relationships have allowed for increased visibility of the varied approaches to using remote sensing to gain a mutual understanding of the archaeological record and the cultures that define that record.

Finally, as the contributors in this volume have all been heavily influenced by the research paths and perspectives initiated by a long list of practitioners (see J. Johnson 2006), we would be remiss if we did not allow for an evaluation of the current state of archaeological remote sensing as presented in the case studies within this volume. We have invited Jay Johnson and Ken Kvamme, two of the many scholars who are pioneers in archaeological remote sensing in North America, to contribute their insights into the development of the discipline of archaeological remote sensing. It is their foundational framework and instruction that have allowed us, and many others, to develop as scholars in the field.

# Acknowledgments

Although our paths crossed only recently with the filming of the PBS TV show, *Time Team America*, practitioners of archaeological remote sensing in North America are an integrated group, to which we have many to thank for the guidance and inspiration needed to compile this edited book. We would like to thank Wendi Schnaufer at the University of Alabama Press for the encouragement to see this book to its completion and to Laurel Anderton for her attention to detail and continuity during the copyediting process.

I (Duncan McKinnon) would like to thank Kent Reilly at Texas State University for his commitment to student success. He granted me the opportunity to explore the world of geophysical survey and develop a passion for the discipline. With Kent's foundation I pursued geospatial studies at the University of Arkansas under Ken Kvamme. As a professor, colleague, and friend, Ken helped me become the researcher I am today, and I am indebted to him for his influence, patience, and guidance. I thank him for all his time and dedication throughout (and beyond) my eight years at the University of Arkansas. Additional thanks go to George Sabo, Jami Lockhart, Jamie Brandon, Chet Walker, Meg Watters, Tim Perttula, and the numerous graduate students and colleagues at various institutions that cannot all be mentioned here but are much remembered and appreciated.

I (Bryan Haley) would like to thank Jay Johnson, who is most responsible for guiding me down the road that lead to my participation in this volume. Jay has an extraordinary ability to recognize important and cutting-edge techniques that are useful to archaeology, to get his students excited about them, and to give them plenty of real-world experience. I'm lucky to have collaborated with him while working on my thesis and in the subsequent eight years while working for the the University of Mississippi Center for Archaeological Research. While at the University of Mississippi, I was also exposed to many of the best practitioners of remote sensing in archaeology, who were important to my professional development but are too numerous to mention here individually. Some that do deserve particular mention include Ken Kvamme, Dean Goodman, Marco Giardino, and Meg Watters. I would also like to thank my colleagues at Coastal Environments, Inc., and at the Tulane University Department of Anthropology.

Finally, we thank the contributors who participated in this volume. It is with their knowledge, interest, patience, and perspectives on the anthropological application of remote sensing that this book was realized. This volume is a product of their contributions to the disciple of remote sensing along with the combined instruction and guidance of numerous mentors, colleagues, and friends who have inspired us all.

Archaeological Remote Sensing in North America

# New Developments in Archaeological Remote Sensing

## An Introduction

*Duncan P. McKinnon and Bryan S. Haley*

*Remote Sensing in Archaeology: An Explicitly North American Perspective*, edited by Jay K. Johnson, was published in 2006 by the University of Alabama Press. It remains a valuable synthesis of archaeological remote sensing that has been the primary guidebook for North American archaeologists interested in applying remote sensing methods. The purpose of the book was to provide a North American perspective on earlier methodological summaries and studies that were largely European based (see Clark 1996; Gaffney and Gater 2003; Scollar et al. 1990; Spoerry 1992) and to outline a series of updated steps or protocols related to data collection, data processing and interpretation, and presentation specific to North American archaeological applications. The book nicely synthesizes many methods that are still in use today and provides an important methodological foundation in the use of remote sensing in North American archaeology.

Our new book coincides with several trends in archaeological remote sensing in North America. The proliferation of these techniques has been paralleled by an increasing focus in journal articles, thematic journal volumes, organized conference symposia, and general session conference papers and posters. For example, an examination of *American Antiquity* from 2006 to 2016 reveals a total of 32 studies that used some type of remote sensing, such as aerial, satellite, or geophysics (for example, see Barrier and Horsley 2014; Burks and Cook 2011; R. Cook and Burks 2011; Davis et al. 2015; Hargrave et al. 2007; J. Harmon et al. 2006; A. King et al. 2011; Kvamme and Ahler 2007; Monaghan and Peebles 2010; Prentiss et al. 2008; Prentiss et al. 2012; V. Thompson and Andrus 2011; V. Thompson et al. 2013; Wright 2014). Published thematic volumes include volumes 29 and 30 of *Southeastern Archaeology* and volume 79 of the *Bulletin of the Texas Archeological Society*. Recently, the international publication *Archaeological Prospection* issued a thematic volume on North American remote sensing (volume 21). Symposia at both regional and national conferences include those at the 2014 Society of American Archaeology meeting titled "Not Just for Specialists—Applying Geophysics in Archaeology," which demonstrated the trend in remote sensing adoption by both academic and cultural resource management (CRM) archaeologists. Finally, the filming of *Time Team America*, an American version of a United Kingdom–based archaeology reality show where remote sensing is prominently featured, is another sign of the increased interest in North American archaeological remote sensing since Johnson's 2006 publication.

In concert with the growth in research and publication has been a significant expansion of technological tools, field methodologies (such as broad-area data collection to address landscape-scale questions), and data integration (such as multi-instrument data collection and fusion). With this expansion of tools comes a suite of new remote sensing approaches in the investigation of archaeological phenomena. These new approaches build upon various uses of the "big four": magnetometry (Kvamme 2006b), resistivity (Somers 2006), ground-penetrating radar (GPR) (Conyers 2006), and electromagnetic induction (Clay 2006; Dalan 2006), and their integration with relevant technologies such as GIS, multi-instrument, GPS-guided, and multisensor arrays. Such integration can facilitate rapid terrestrial-based data collection over broad areas, three-dimensional subsurface data visualization with GPR and electrical resistivity tomography, and advances in airborne (LiDAR) and satellite remote sensing approaches.

Accompanying the increase in North American remote sensing technologies and methods are the integration of anthropological theory and the development of interpretations beyond the application of simple map-based prospection and the location of archaeological features. Today, remote sensing data are frequently utilized as a primary data set that can be integrated into larger research programs associated with investigating and evaluating anthropological attributes of the cultural past (Aspinall et al. 2008; Kvamme 2003a; V. Thompson et al. 2011). This action encourages the use of archaeological remote sensing data as a tool directly involved in interpreting change and continuity of culturally constructed environments as "a departure point to discuss broader implications related to social and cultural meaning" (V. Thompson et al. 2011). Stated differently, "prospection alone need not be the ultimate goal . . . even more exciting is the use of [geophysical] results to help explain aspects of ancient cultures that can be known in no other way" (Aspinall et al. 2008:45).

Students of North American archaeology are increasingly interested in incorporating anthropological remote sensing into their graduate research. Additionally, professors of "traditional dirt archaeology" have become more engaged in procuring instrumentation to address research problems—not only for their students, but also to expand their own objectives. Furthermore, field schools at universities throughout North America are incorporating remote sensing into their instruction. For example, the Center for American Archeology (CAA) has integrated remote sensing into its annual archaeology field school with instruction in remote sensing methods, data processing, and theory-based interpretation (see chapter 11). The CAA field school focuses on a student-centered approach where students gain empirical and theoretical training in research design and implementation embedded within a regional long-term program (Buikstra et al. 2011; McKinnon et al. 2016).

Trends in academia have predictably been mirrored by trends in CRM archaeology, where the ability of remote sensing techniques to efficiently characterize buried features is at a premium. Remote sensing techniques are encouraged by most, if not all, guidelines and standards for professional archaeology. Given the temporal and financial demands of many CRM projects, remote sensing data have the ability to enhance the anthropological conclusions of these projects.

With the expansion of remote sensing in North American archaeology, students and practitioners need current exemplary applications that adhere to high academic standards in terms of methodology, emerging technologies, data processing and presentation, and theory-based interpretation that are specific to North American archaeology. Our volume provides a series of case studies to address this need.

The integration of theory into archaeological remote sensing demonstrates that quantitative and qualitative approaches in evaluating cultural landscapes need not be independent of each other. The complementary use of both quantitative and qualitative approaches allows for an examination of the integrated functional and symbolic components that more accurately define complexities of past cultures (Blitz 2010; McKinnon 2013b; Sabo 2012; V. Thompson et al. 2011). With this combined approach, cultural landscapes can be considered holistic in nature, embedded with social or cosmological meaning, purpose, and vision and integrated into the daily lives and functions of occupants (Anschuetz et al. 2001; Cobb 2003; Kornfeld and Osborn 2003; Robb 1998; Sabo 2012). These integrated functional and symbolic components are often manifested as a multitude of spatial resolutions and relationships, including regional-level relationships to natural and culturally constructed features and other village locales (Binford 1982; Blitz and Lorenz 2006; Girard 2012; Lockhart 2012; Marrinan and White 2007; Perttula 2009; Pollack 2004; Rafferty and Peacock 2008; Rees and Livingood 2007; Sabo 2008; Vogel 2012), intrasite or intraorganizational scales reflected as the deliberate spatial orientation of special-use structures, activity areas, and delineated spaces (Bailey 1995; J. Brown 2012; Haley 2014a; Henry 2011; A. King et al. 2011; Knight 1998; Lockhart 2007; Perttula and Rogers 2007, 2012; Sabo 2012; Walker 2009; Walker and McKinnon 2012), and small-scale resolutions examining differential use and spatial organization, such as on mound summits or within individual households, and the use of interior space (Deetz 1982; Gasco 1992; Jackson et al. 2012; A. King et al. 2011; Pluckhahn 2010; Schambach 1996; Trubitt 2009; Wilk and Rathje 1982).

Recent geophysical explorations have continued to investigate space and place in terms of the traditional larger social and environmental processes, but also with a newfound attempt at understanding individual contributions to the spatial makeup of "persistent places" over time (Schlanger 1992: 92; for examples, see Aspinall et al. 2008:245; A. King et al. 2011; Perttula et al. 2008). These new landscape-level characterizations of the spatial components of places are allowing archaeologists a more informed understanding of the social and cultural factors related to architectural change over time, demarcations and definitions that constitute landscape use, and continuities of architectural and related cultural traditions across space and time.

In line with these developing perspectives, *Archaeological Remote Sensing in North America: Innovative Techniques for Anthropological Applications* builds upon the methodological foundation established in *Remote Sensing in Archaeology: An Explicitly North American Perspective* and presents a series of case studies from scholars throughout North America who are utilizing archaeological remote sensing in progressive ways to address anthropological problems. Contributions are geographically diverse (Figure I.1) and include the Southern Plains (Hammerstedt et al., chapter 1), the mid-South (Patch et al., chapter 2; McKinnon and Haley, chapter 3; Zimmer-Dauphinee, chapter 13), the Midwest (Henry et al., chapter 10; King et al., chapter 11), Northern Plains (Wiewel, chapter 6; Lynch and Becker, chapter 9), Florida Gulf Coast (Thompson, chapter 8; Bigman and Seinfeld, chapter 12), Atlantic Coast (Lynch and Becker, chapter 9), upper Midwest (Lynch and Becker, chapter 9), and three distinct regions in the Southwest (Mink, chapter 4; Sturm, chapter 5; Diederichs et al., chapter 7).

Unlike the 2006 volume, which was organized around specific techniques, this book is organized around four major anthropological themes: Site Structure and Community Organization, Technological Transformation and Economic Change, Archaeological Landscapes, and Earthen Mound Construction and Composition. Within each section, contributions illustrate and emphasize new and informative applications using theory-based archaeological remote sensing interpretation. For example, work at Spiro (chapter 1), at sites along the Tennessee River Valley (chapter 2), at the

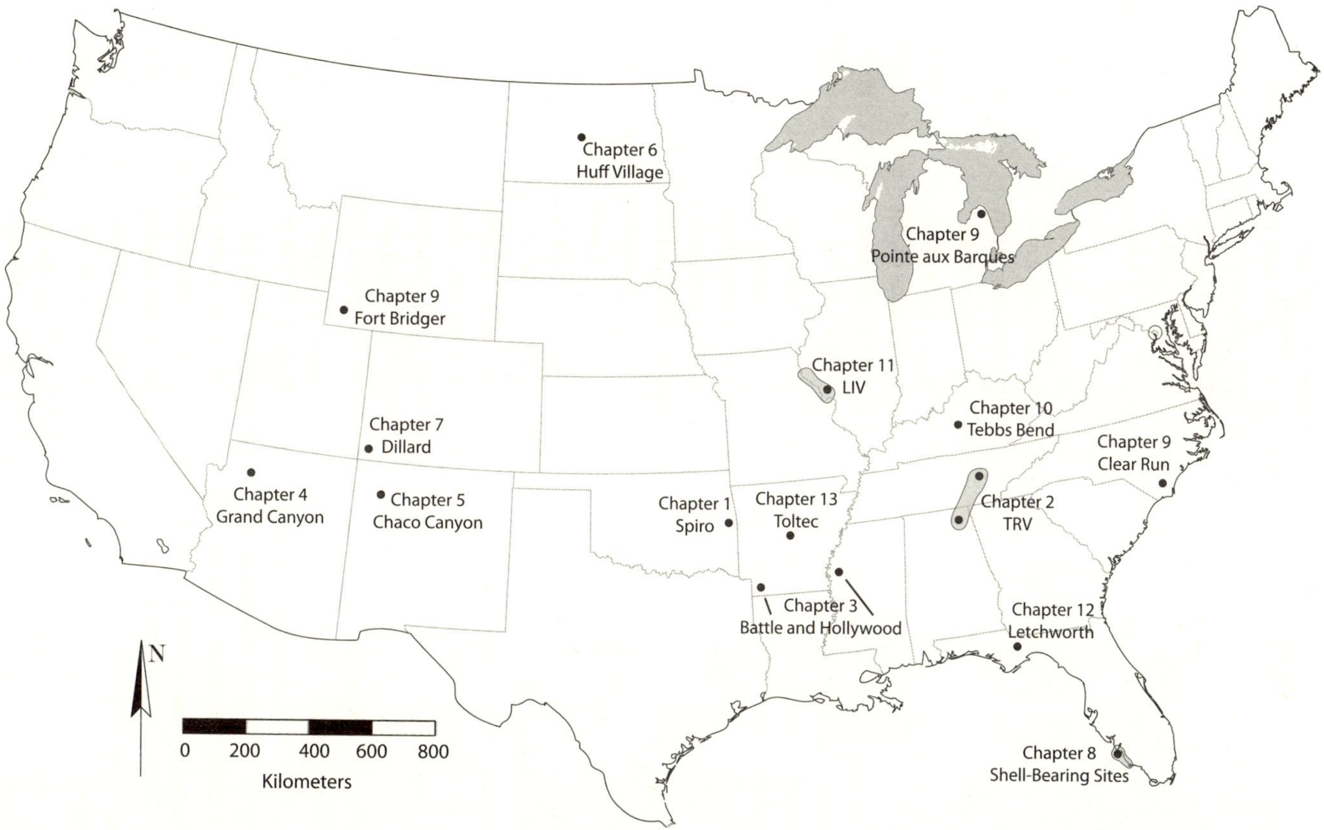


Figure I.1. Distribution of sites and regions discussed in the text.

mound centers of Battle and Hollywood (chapter 3), and at several sites within Grand Canyon National Park (chapter 4) all focus on the application of archaeological remote sensing in evaluating the distribution of anomalies across sites, inferring the distribution of archaeological features from those anomalies, and determining how the resulting insights into site structure inform anthropological themes such as community organization. Research at Chaco Canyon (chapter 5), Huff Village in North Dakota (chapter 6), and the Dillard site in the Four Corners region (chapter 7) are focused on using specific geophysical and archaeological features (agricultural fields, pits, and kiva/pueblo architecture) to evaluate and understand technological and economic processes related to cultural change over time. Chapters 8, 9, and 10 broaden the scope of inquiry by utilizing archaeological remote sensing to examine cultural landscapes and their use and modification, and, where possible, to gain insight into concepts of meaning embedded within archaeological landscapes. Finally, research at sites containing Woodland period earthen mounds within the Lower Illinois River Valley (chapter 11), on the Florida Panhandle (chapter 12), and along the Arkansas River in central Arkansas (chapter 13) provide important case studies using archaeological remote sensing to map, decipher, and interpret cultural traditions related to mound construction and internal composition. Chapters 11, 12, and 13 demonstrate the value of noninvasive methods in visualizing the processes of monumental construction in situations where excavation is rarely employed today for a variety of methodological, economic, and ethical reasons.

The diversity in data presentation and coverage provided by the chapters in this book is a real strength and a testimony to the application of archaeological remote sensing in numerous environments and under varying conditions (see Kvamme 2006a, 2006b). In other words, these are "real" data and not a suite of pick-and-choose case studies where there is a "tendency for geophysicists to distribute images from their most dramatic surveys" (Hargrave 2006:269). In the real world, archaeologists have to deal with nonideal circumstances and produce and interpret results to address real research questions. These real-world expectations are relevant in a variety of archaeological applications, including CRM and academic research where time or resources are limited.

Overall, the distribution of case studies in *Archaeological Remote Sensing in North America* reflects multiple geographic environments and related archaeological regions that make up a large part of North America. With this in mind, we emphasize the temporal diversity of the book (Figure I.2). Case studies represent a significant amount of time depth. Chronology ranges from the early Florida Gulf Coast shell-bearing sites discussed in chapter 8 to the nineteenth-century historical archaeology discussed in chapters 9

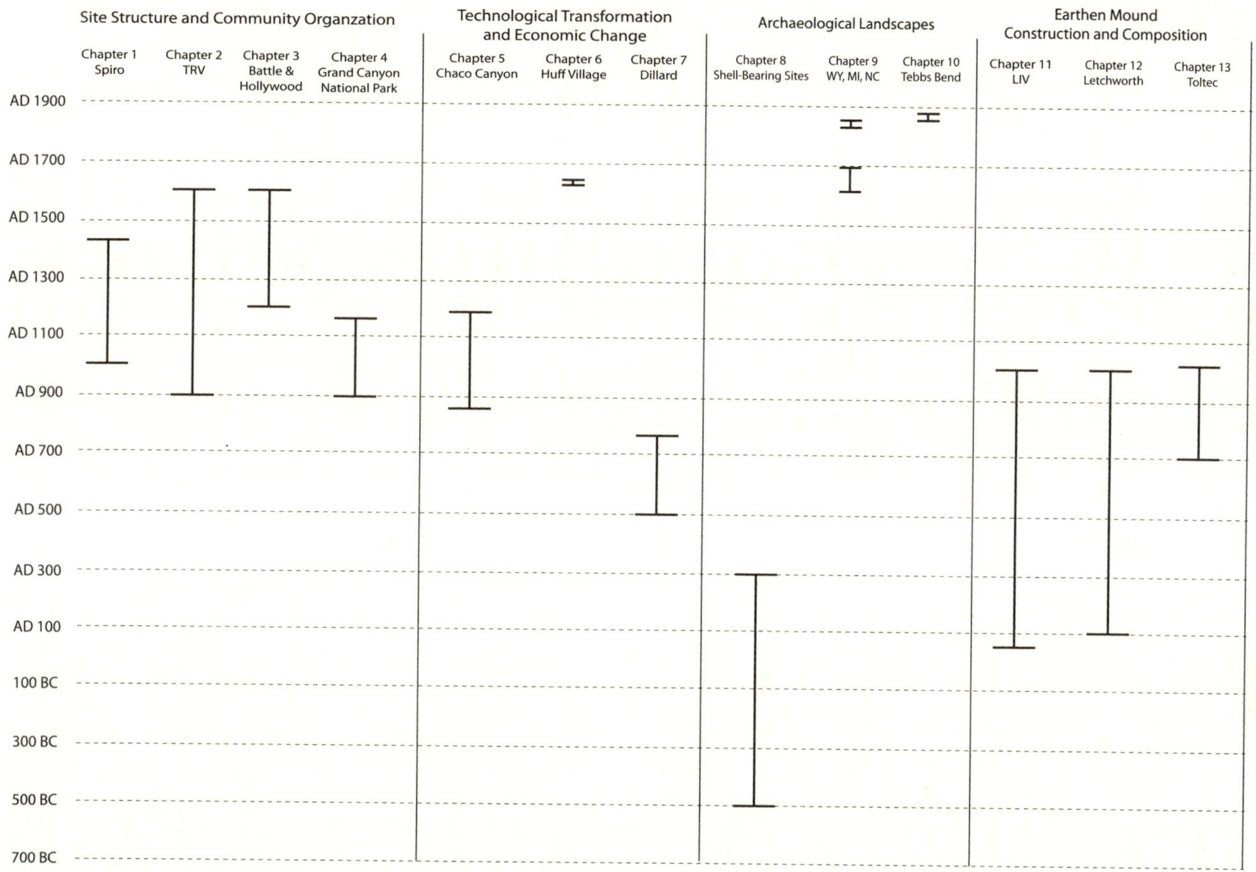


Figure I.2. Temporal association of sites and regions discussed in the text.

and 10. The particular distribution of case studies in this book should not be viewed as an absolute reflection of where remote sensing is being used at the greatest frequencies. It is probably more a reflection of our professional networks and those of our colleagues, although we did attempt to be all-inclusive.

Finally, we have chosen to avoid dwelling on differences in terminology used by the various contributors and to let them develop organically. As was the case with the 2006 volume, we have used the term "remote sensing" in the title. With the exception of chapter 9, which uses LiDAR data collected from an aerial platform, all the case studies rely mostly on "geophysics," which can be understood as a special type of remote sensing that is performed on the ground surface. Of course, geophysical techniques collect information on physical properties under the ground surface, while remote sensing from aerial platforms generally responds only to the ground surface. While we recognize this difference and potential confusion, our choice is based on the fact that "remote sensing" resonates best within the archaeological community, as well as the fact that airborne and near-surface applications share certain similarities related to processing and data structure. Additionally, several contributors use the term "ground-truthing" in reference to excavated data sets that serve to augment remote sensing data. As Hargrave states (2006:269), "*truthing* refers to the interpretation of the remote sensing data; it does not imply that the actual data may be spurious." Importantly, the term is frequently used in archaeological literature, such that archaeologists, especially those doing CRM, use "ground-truthing" primarily when referring to excavations or other forms of invasive data collection.

# I

# Site Structure and Community Organization

# 1

# Multisensor Remote Sensing at Spiro

## Discovering Intrasite Organization

*Scott W. Hammerstedt, Jami J. Lockhart, Patrick C. Livingood, Tim Mulvihill, Amanda L. Regnier, George Sabo III, and John R. Samuelsen*

The Spiro Mound Group in eastern Oklahoma (see Figure I.1) is best known for the remarkable objects and ritual tableaus that have enhanced the study of Mississippian iconography and exchange. However, with few exceptions, excavations and collection research have focused on mounds and mound deposits. In this chapter, we integrate evidence from a site-wide geophysical survey (using a gradiometer, electrical resistance, electromagnetic conductivity, and magnetic susceptibility) and fine-grained mapping to better understand off-mound occupation and reconstruct the Spiro landscape. Our results provide a better picture of site use, complement existing data from previous excavations, and enable us to locate archaeological deposits that would not have been discovered using traditional archaeological methods.

## Spiro

Spiro was occupied from at least A.D. 800 to 1400 (Figure 1.1) and was the largest, and probably most influential, mound site in this portion of the Arkansas River drainage. It is currently maintained by the Oklahoma Historical Society as an archaeological park. Spiro can be divided into two main areas based on topography; these are referred to as the upper and lower terraces (Figure 1.2). The multilobed Craig Mound, which contained the Great Mortuary, is located on the lower terrace along with the two smaller Ward Mounds. On the upper terrace is a series of mounds arranged around a possible plaza. Brown and Copple mounds are relatively large platforms and the remainder are buried house or midden mounds.

Rogers (2011) described Spiro as having a tripartite division of space. This division began during the Harlan phase, when Spiro started its rise to prominence, and continued into the subsequent Norman and Spiro phases,

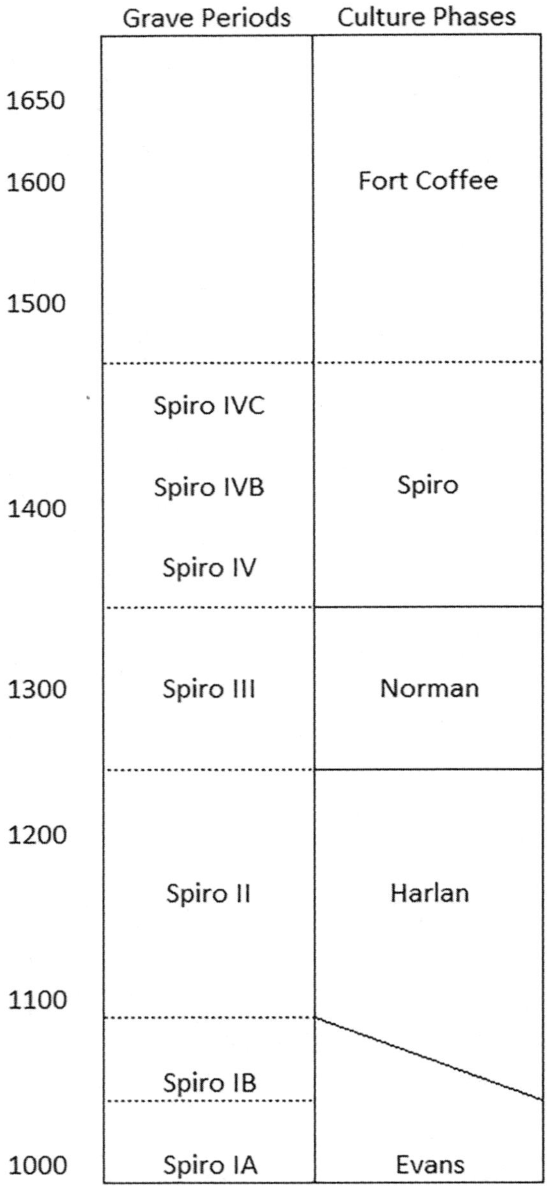


Figure 1.1. The Spiro sequence (after J. Brown 1996; Rogers 2011).

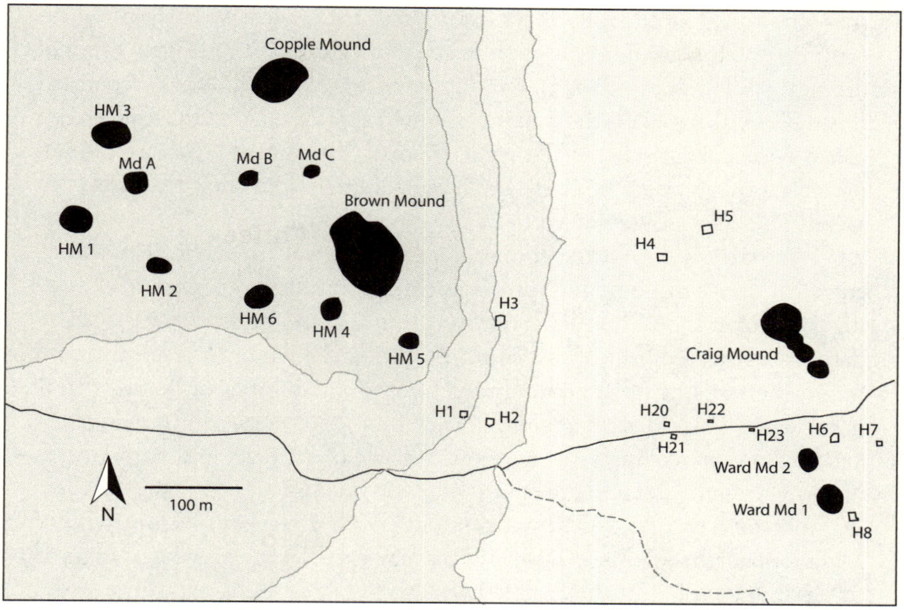


Figure 1.2. Map of Spiro showing the locations of mounds and excavated lower terrace buildings. The solid line marks the path of a creek; the dashed line shows the pre-1940s path. Md, mound; H, house; HM, house mound.

when Spiro was at its most culturally complex. The westernmost division, on the upper terrace, consists of platform mounds (used in part for mortuary activities), house mounds (created by periodic rebuilding of structures) that contained special-purpose buildings, and midden mounds that may have been created by occasional cleaning of the plaza. The central part of the site extends from Brown Mound down the slope to the lower terrace, and east to Craig Mound. Excavations conducted by the Works Progress Administration (WPA) encountered eight structures in this area; these will be discussed further below. The third area, represented by Craig and Ward mounds, is related primarily to mortuary activities that culminated with the Great Mortuary.

Spiro is world renowned for the spectacular artifacts found within Craig Mound. These include engraved marine shells, copper, textiles, pottery, shell beads, wooden masks, cedar litters, and effigy pipes imported from as far away as the Gulf Coast, Great Lakes, Mississippi Valley, Mexico, and California (Barker et al. 2002; J. Brown 1996; Hamilton 1952; Kozuch 2002; Sievert and Rogers 2011). James Brown (2012) recently reinterpreted the Great Mortuary as a cosmographic tableau created as part of a ritual event that drew in people from the surrounding region in the early fifteenth century, possibly as an attempt to ward off an extended multigenerational drought (E. Cook et al. 2007). Even more recently, Brown (2014) separated this tableau, situated on the floor of a domed structure, from the underlying Great Mortuary ossuary/cedar litter deposit. The domed structure (which preserved the famous "hollow chamber") along with its enclosed display of sacred objects is now referred to as the Spirit Lodge.

Over the last 100 years, sporadic archaeological excavations have occurred at Spiro, through both professional excavation and commercial looting (J. Brown 1996). Joseph Thoburn of the Oklahoma Historical Society partially excavated both Ward Mounds during the winter of 1916–1917 and found the remains of submound structures and a number of burials (Thoburn 1931). Craig Mound, including the Great Mortuary, was heavily damaged by looters seeking to profit from the sale of artifacts in the mid-1930s (J. Brown 1996; Clements 1945; La Vere 2007).

From 1936 to 1939, WPA workers under the direction of University of Oklahoma archaeologists conducted salvage excavations on what remained of Craig Mound after the looters had finished (Clements 1945). The WPA also tested many of the remaining mounds, expanded on Thoburn's work at Ward Mounds, and excavated 14 structures across the site (J. Brown 1996; Orr 1946).

More recently, crews directed by Don Wyckoff of the Oklahoma Archeological Survey relocated the footprints of Craig and Ward mounds to allow their reconstruction as part of the site's conversion to a state park (Wyckoff 1970). From 1979 to 1982, Oklahoma Archeological Survey crews led by J. Daniel Rogers tested several house mounds (most notably House Mound 5), the Brown and Copple platform mounds, and four midden mounds. Rogers's team also conducted limited off-mound testing and a small-scale magnetometer survey (Rogers 1980, 1982a, 1982b).

While each of these projects contributed valuable data, they were necessarily limited in scale by Spiro's size and the archaeological and geophysical techniques available at the time. Further, mounds (both platform and house) were prioritized throughout the excavation history. As a result, we have a working knowledge of the construction and use sequences of many of the mounds but know very little about the extensive nonmound areas. Thus, our picture of Spiro's history is skewed.

What little we know about nonmound occupation comes largely from WPA work. WPA crews excavated all or part of 14 structures. Eight of these were not associated with a mound and were on the lower terrace, and six were within mounds on the upper terrace. The off-mound structures were identified by the presence of artifacts, especially clay daub, on the plowed surface of the lower terrace. Nearly all of these structures are square or rectangular, with single-set wall posts and four center support posts. Of the eight nonmound structures, four have neatly spaced straight walls and four central support posts, and four have relatively sloppy wall post patterns. Of those with sloppy wall post patterns, three have coherent patterns of four center posts, and excavation of the fourth was incomplete. House 2 has two dates, one obtained recently; when pooled, these dates have a 1-sigma calibrated age of A.D. 1165 to 1219 (O-2262; Beta 388537). House 4 has a calibrated 1-sigma range between A.D. 900 and 1150 (Tx-3611). Both dates put these structures in the Evans or Harlan phases, at the earlier end of Spiro's sequence (J. Brown 1996:123–132). However, the lack of datable material and the lack of spatial coverage make any extrapolation of these data to other parts of the site problematic.

## The Spiro Landscape Archaeological Project

With the above concerns in mind, we launched the Spiro Landscape Archaeological Project (SLAP) in 2011. The initial research goals of the project are to obtain much-needed chronological and spatial data on Spiro's nonmound areas and to conduct a landscape geophysical survey of the entire landform that surrounds the site. The data will complement information gleaned from decades of research on Great Mortuary deposits and integrate the ritual and domestic landscape at and around Spiro itself. To this end, we conducted a gradiometer survey of all nonforested areas of the site to provide full coverage of nonmound areas. Historical aerial photos dating to the 1930s have been useful in reconstructing Spiro's natural and cultural landscape. The original course of a small drainage that cuts across the lower terrace is visible in the earlier photos. This is compared to the human-made channel (1940s) that now trends due east and west, artificially dividing the site (see Figure 1.2). LiDAR is not yet available, but we used a robotic total station to collect close-interval elevation points, which we interpolated for accurate on-site topography.

Using a total station, we set up a system of 20 x 20 m grids on the previously established site coordinate system to enable comparison with earlier archaeological work. The geographic coordinates of grid corners were collected with a Trimble GPS receiver and differentially corrected so that all the work is now also georeferenced in UTM NAD83. All the data described here have been entered into a geographic information system (GIS), so they can be used with a wide variety of background data layers that we continue to assemble for the project.

Gradiometer data were collected using three Bartington Grad601 gradiometers and one Geoscan Research FM256. It took five weeks in the field over a three-year period to collect the initial 30 ha of gradiometry data. Additional time was needed for digital processing, concatenation, and georeferencing. The field resolution was 50 cm in the *X* direction (east-west) and 12.5 cm in the *Y* direction (north-south). The data were interpolated to 12.5 cm cells for the entire area. Based on the gradiometer results, selected areas were also successfully surveyed using electrical resistance, electromagnetic conductivity, and magnetic susceptibility. Ground-penetrating radar was also employed but with much less success; it is not discussed here, although it has been previously used at Spiro by others (Goodman 2004; Goodman et al. 1995).

## Remote Sensing Results

Gradiometer results show that the surface of the site has been subjected to a range of natural and cultural disturbances. Initial interpretations involved identifying and making distinctions between various sources of land alteration. One immediately apparent pattern was a change in data texture on the lower terrace along a line trending northwest to southeast and encompassing the most easterly quarter of our data set (Figure 1.3). We initially suspected mechanized disturbance. However, when soil boundaries were overlaid onto the gradiometry, it became apparent that this change was due predominantly to a change in soil type from a Redport silty clay loam to a Coushatta silt loam (Abernathy et al. 1983), which precisely matches the boundary of the magnetic data trend. That soil change is related to the highly dynamic character of the natural landscape near the site. Meander scars seen in historical photographs show that the Arkansas River once flowed immediately adjacent to the large Craig Mound, although it is not yet clear whether the river was in this location at the same time the mound was in use. This potential juxtaposition between the primary mound and the river may have implications for intrasite organization, as the pattern of large mounds being intentionally placed at river's edge is seen at numerous Mississippian sites throughout the southeastern United States. More-focused geomorphological studies are planned to examine this intriguing potential.

### Upper Terrace

Our gradiometry survey reinforces the notion that the upper and lower terraces at Spiro are fundamentally different (Rogers 2011). Before our work, some of these differences were readily apparent. The upper terrace consists of a ring of mounds around an apparently open plaza, while the lower terrace shows no detectable patterning of mounds. Our gradiometer survey

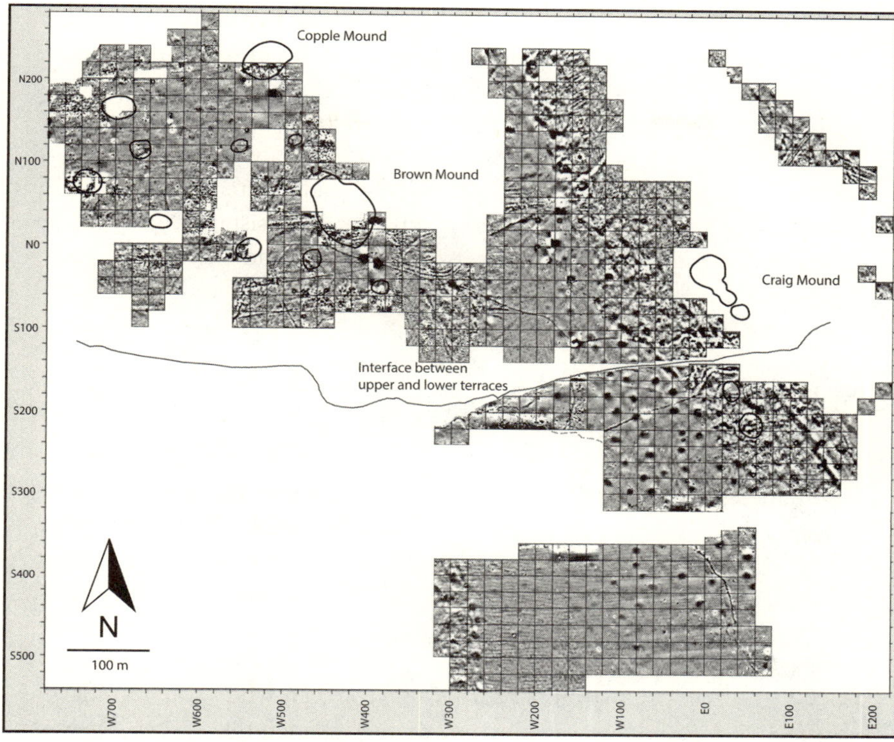


Figure 1.3. Gradiometer results.

identified significant additional differences between these areas not otherwise apparent in the visible built environment or known archaeological record.

Numerous small magnetic dipoles characteristic of modern historical metal detritus are present (Figure 1.4a). This is hardly surprising given that a number of tenant farmers maintained houses and farms on the upper terrace until the 1960s. However, other magnetic anomalies have signatures consistent with prehistoric features (2–4 nT). There are a number of square and rectilinear anomalies on the upper terrace; one example of these is a 6–7 m anomaly shown in Figure 1.4b. As noted, previously excavated buildings at Spiro are square or rectangular. Therefore, we have interpreted this anomaly as a building. In another area of the upper terrace, the gradiometer results show the relationship between discrete areas of enhanced magnetic strength and large and small topographic rises (see Figure 1.4a). Some of the more obvious rises were tested during previous excavations. The larg-

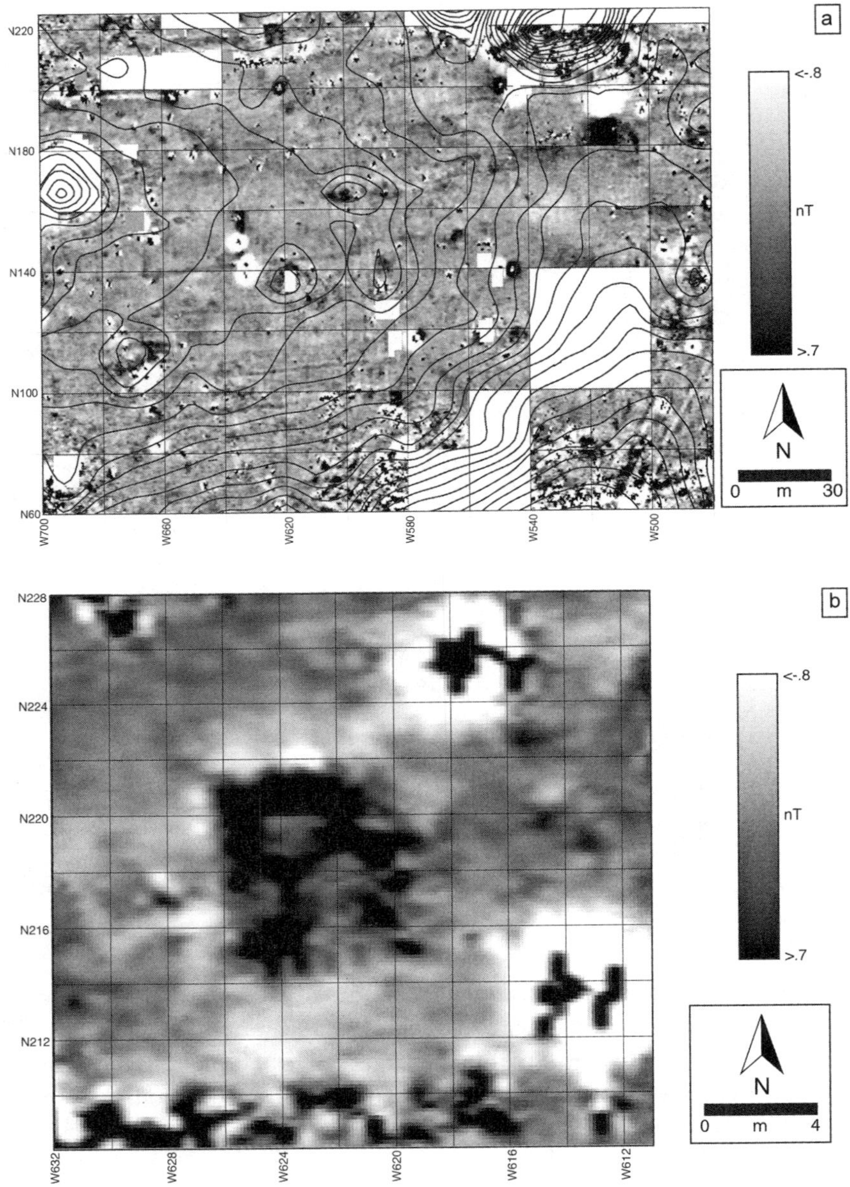


Figure 1.4. (a) Portion of the upper terrace showing historical debris and areas of enhanced magnetic strength on small topographic rises. (b) Close-up of a probable square building, which can also be seen on the top center of (a).

est rises, the house mounds, consist of dismantled or burned structures deliberately covered with earth. Other, smaller rises have proven to be midden mounds that likely date to the Harlan phase (J. Brown 1996:115–122; Rogers 1982a, 1982b). Our close-interval topographic mapping and gradiometer survey demonstrate that a number of subtle rises with associated amorphous areas of increased magnetism are present on the upper terrace. None of these have been excavated; we hope to test several of these features in the future.

Given the arrangement of the upper terrace mounds in a loose circle, it has been hypothesized that the enclosed area was a plaza (e.g., Rogers 1982b). The results of our gradiometer survey reinforce the notion that it may have been one for at least part of Spiro's history. The gradiometer data suggest that a high-density residential area does not exist in this part of the site. Based on current data, it appears buildings were present on the upper terrace, but at a low density. The presence of these buildings, along with the small midden mounds, indicates that while it may have been a plaza at some point, it was not always vacant space. Excavation could provide valuable data concerning the position of the presumed buildings and midden heaps in the existing site chronology and help us better understand whether they were built for ritual or domestic use.

### Interface between Terraces

In the area of steep ground dividing the upper and lower terraces, we identified a row of regularly spaced, nonmetallic, monopolar magnetic anomalies with a signal strength consistent with large burned features (some up to 9 nT). These appear to be smaller than structures and may represent a symbolic but structural division between the upper and lower terraces (see Figure 1.3).

### Lower Terrace

The lower terrace exhibits a different layout. A pattern of numerous discrete, structure-sized magnetic anomalies is apparent (Figure 1.5). These anomalies are barely visible within the larger range of background magnetic signal strength—some less than one nanotesla above background. Consequently, Figure 1.5 has been heavily clipped to make these subtle anomalies more visible.

The presence of these anomalies was initially surprising because this was believed to be a plaza or otherwise unoccupied area, and therefore not of great archaeological interest. The anomalies are 6–10 m in diameter and are roughly circular, making them the right size and shape to be structures. At least 63 are visible in the gradiometer results. They are relatively evenly spaced across the landscape, with no evidence of overlapping rebuilding epi-

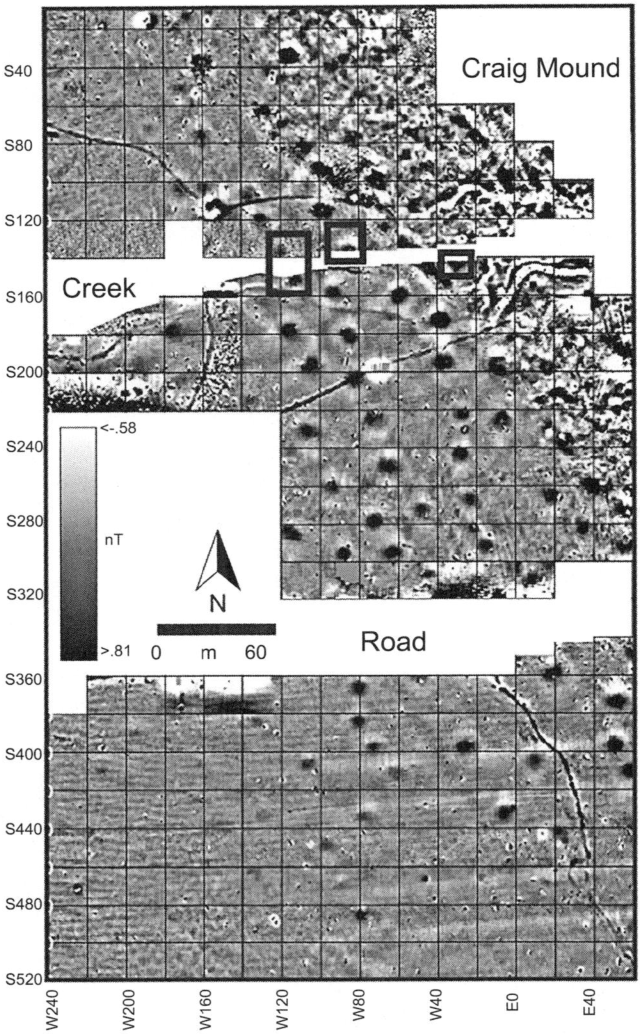


Figure 1.5. Lower terrace showing numerous structure-sized magnetic anomalies. Craig Mound is located in the upper right.

sodes. Further, some of the anomalies form rows oriented in approximately the same semicardinal direction as Craig Mound. The even spacing, consistent shape, lack of evidence of rebuilding, and weak magnetic signature are all consistent with a simultaneous construction of dozens of short-term structures on the lower terrace.

Because of the low magnetic signature of these anomalies, we were initially cautious with our interpretation of them as intentional structures. However, the use of multisensor geophysics has corroborated these magnetic anomalies (Figure 1.6). In addition to gradiometry, we have used electrical resistance, electromagnetic conductivity, and magnetic susceptibility. All have been useful in providing insights for specific feature types. In electrical resistance data, the features are often characterized by a rounded and centered area of decreased resistance ringed by increased resistance readings, indicating there is slightly more moisture retention within them. Magnetic susceptibility has also been very useful in corroborating these subtle, structure-like magnetic anomalies. The technology is also showing promise for high-resolution work on soil profiles to identify the contemporaneous living surface.

In addition to these ephemeral features, a number of square anomalies are evident near the interface between the upper and lower terraces. These have a stronger magnetic signature that suggests they are more like the buildings excavated in this area by the WPA. As discussed, the available radiocarbon dates place two of these buildings in the Harlan phase. It is possible that there may have been a Harlan phase occupation of this portion of the lower terrace.

During the fall of 2011 and summer of 2014, we used a Geoscan Research RM15 electrical resistance meter with multiplexer to investigate a large contiguous area (roughly 1 ha) south and east of Ward Mounds. This area is within the meander scar soil discussed above and was specifically selected to test the suitability of different technologies in areas in which gradiometry was not oversuccessful. The resistivity survey showed that (1) it is more suited to work in this soil type than gradiometry; and (2) structures not seen in the gradiometer data are likely present (Figure 1.7). Although their exact locations are not clear, the WPA-excavated House 8 lies within our survey area, and Houses 6 and 7 are located just to the north (see Figure 1.2). A reconstruction of House 6 currently stands in its estimated location. An anomaly of the same shape, size, and orientation as House 8 is present near its possible location (see Figure 1.7).

Rogers (1980, 1982a) excavated a number of 1 x 1 m test units, postholes, and a Gradall trench 40 m long, 0.9 m wide, and 30 cm deep in this area. The trench was visible in both resistivity and gradiometry and is represented by a solid black line in Figure 1.7. This line intersects the edge of a structure-like anomaly. According to Rogers (1980:35), the trench produced few artifacts and no features were identified. However, it is possible that any structures are deeper than the trench, or they are difficult to detect since the trench was narrow. Further, some of the areas of high resistivity may correspond with use areas rather than structure boundaries.

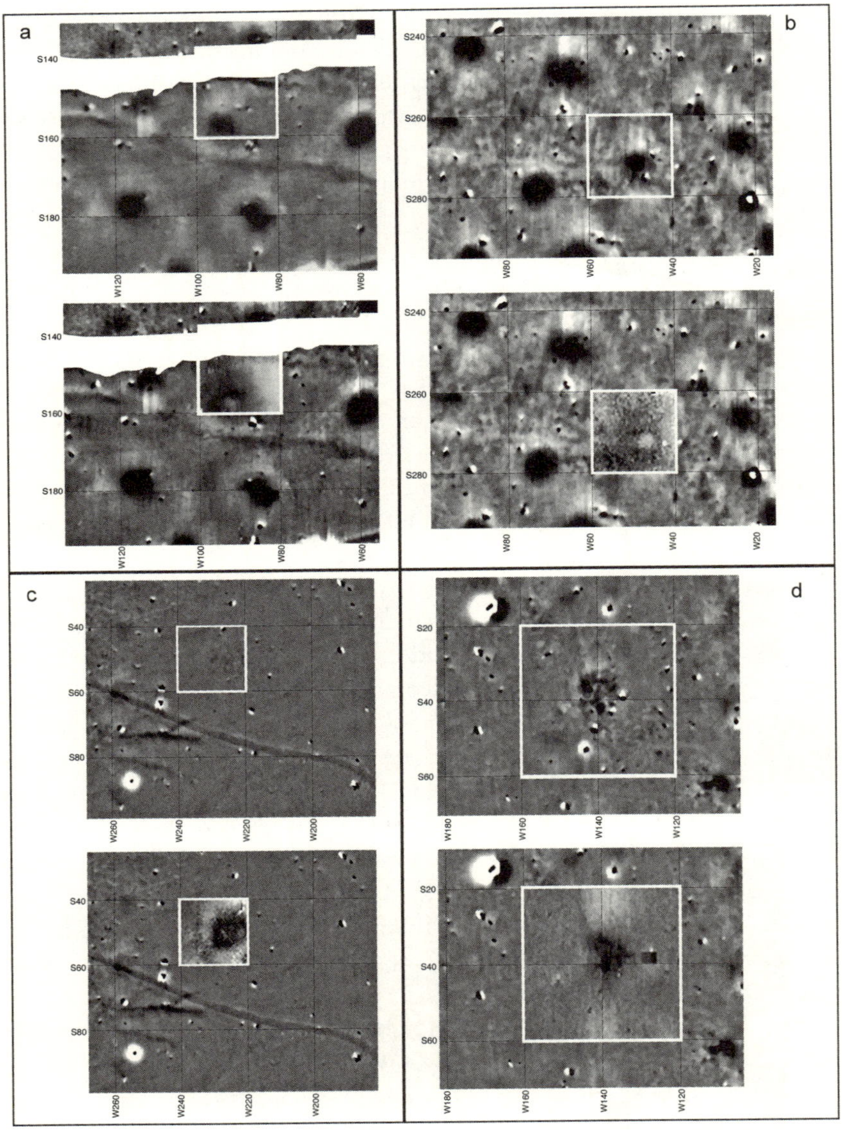


Figure 1.6. Comparison of different technologies depicting the same anomalies: (a–c) gradiometry (top), resistivity (bottom); (d) gradiometry (top), magnetic susceptibility (bottom). Refer to Figure 1.5 for location on the site. Grids are 20 m.

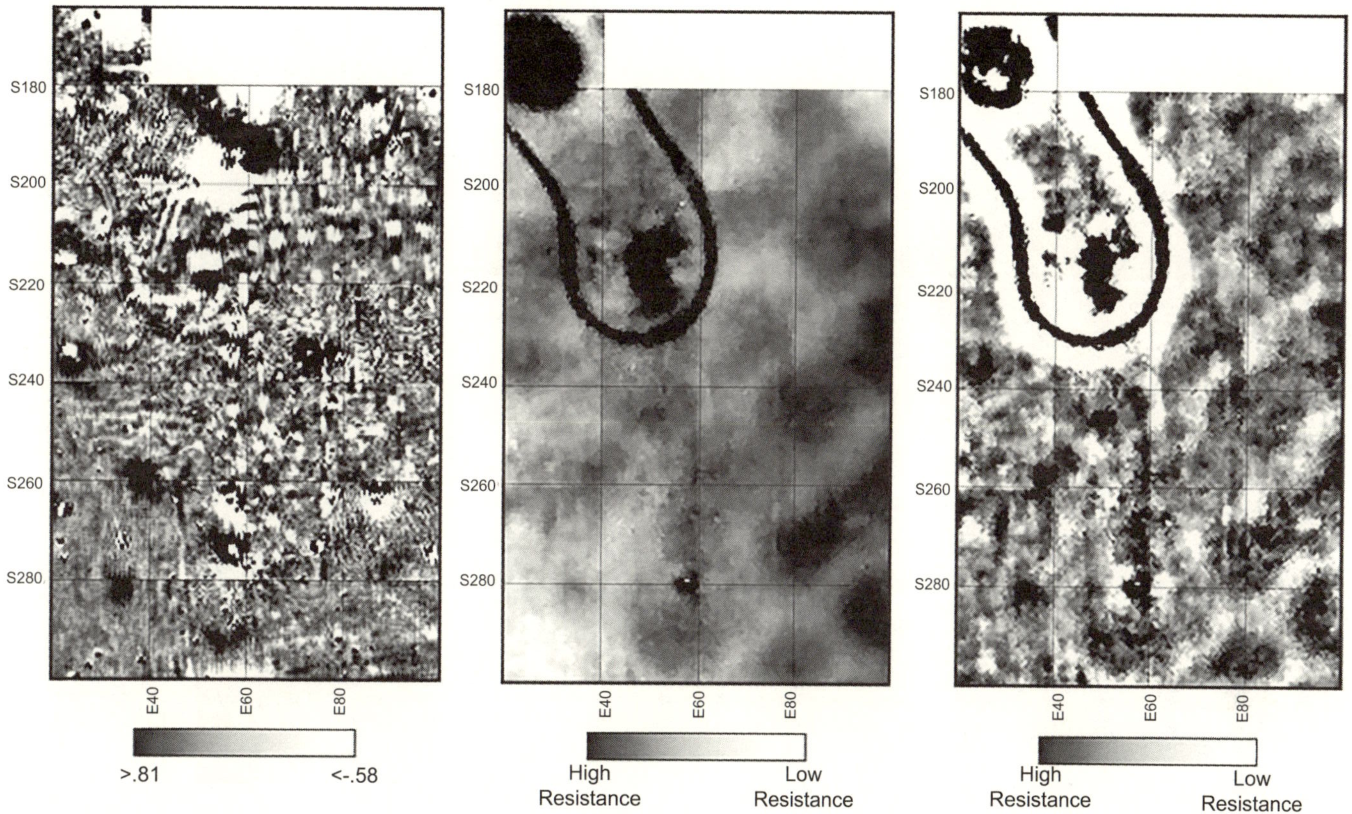


Figure 1.7. Anomalies near Ward Mounds: left, gradiometer; center, resistance; right, resistance with a high-pass filter applied. The linear band to the top left is a gravel path encircling the mounds. See Figure 1.5 for location on the site. Grids are 20 m.

Resistivity also identified some areas of known historical disturbance. A north-south road can be seen in a 1938 aerial photograph (J. Brown 1996:14, Figures 1–5) passing through the east side of Ward Mound 1. This is also visible in the data as a long north-south linear anomaly. While the gradiometry data show small squares (2.5 x 2.5 m) of low magnetism, it is unclear what they represent, as they do not appear in resistivity or any of the other technologies. It is possible that they represent WPA or other trenches that are not well documented.

### Summary

In total, our gradiometer survey covered 30 ha, with selected grids within that area surveyed with multiple technologies. The results suggest that our use of multiple technologies was crucial to our interpretation of the data, a point made long ago by others (e.g., Clay 2001). The full-coverage survey supported some previous theories about site use, while also revealing the presence of a number of unexpected anomalies indicative of buildings near Craig Mound.

## Ground-Truthing

Four of the structure-like anomalies identified during the gradiometer survey were along an artificial stream cut across the lower terrace by a resident farmer during the 1940s (see Figure 1.5). Erosion along the stream has been an ongoing problem, most dramatically illustrated by a US Army Corps of Engineers boundary marker that has largely washed out of the stream bank over the past few decades. Because of the threat to these four anomalies, we obtained permission from the Corps of Engineers, in cooperation with the Caddo Nation of Oklahoma, the Wichita and Affiliated Tribes, and the Oklahoma State Historic Preservation Office to conduct archaeological salvage excavations. Two were south of the creek, and two were on the north side. Following our agreement with the Corps, we did not excavate within 1 m of the stream edge to keep from further worsening the erosion.

Fieldwork took place in the fall of 2013 and the spring/summer of 2014. In each case, archaeological features confirmed the gradiometer and resistivity results (Figure 1.8) and suggest that these anomalies were prehistoric structures. Post molds and postholes corresponded very well with the boundaries seen in the geophysical data, and we have designated these as Houses 20–23 in keeping with the numbering system used by previous researchers (J. Brown 1996; Rogers 1980, 1982a, 1982b; Rogers et al. 1989). Analyses are ongoing, but we are able to make some summary statements here: post patterns were haphazard (not quite square, not quite circular) but

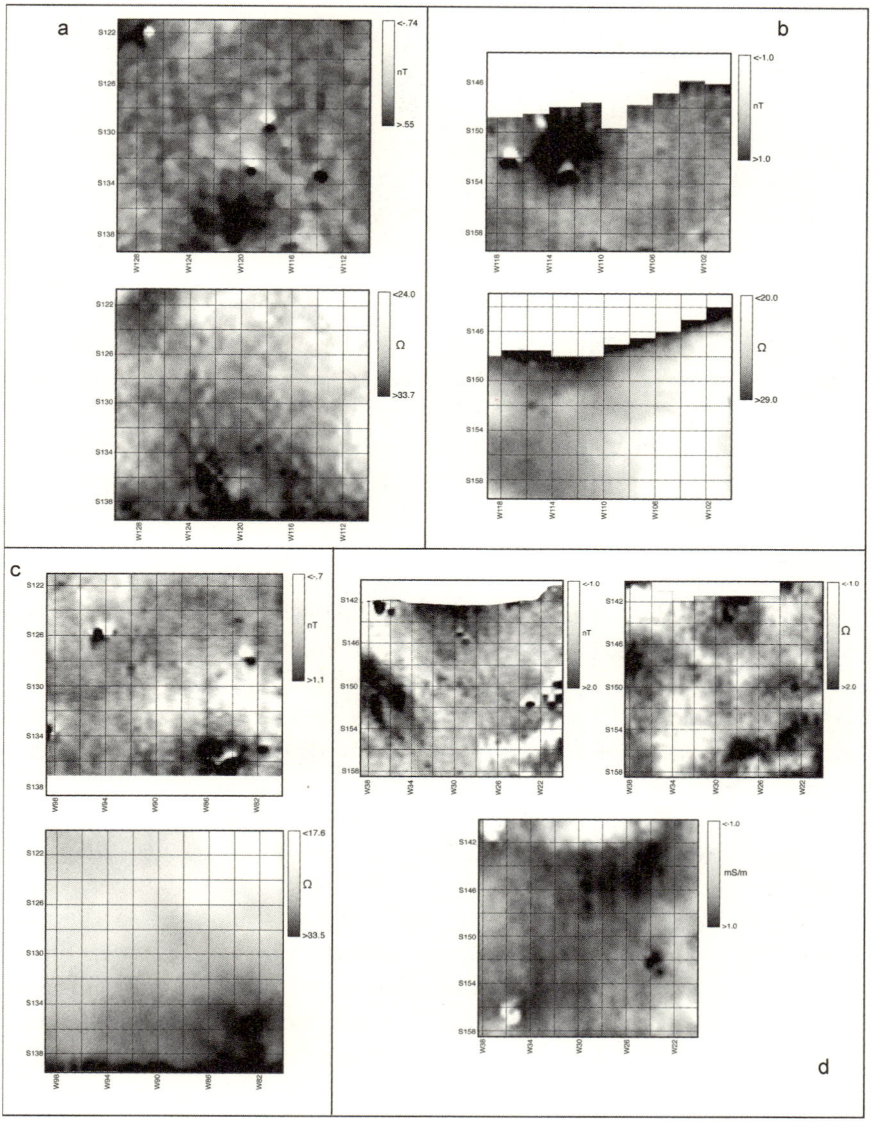


Figure 1.8. (a) House 20; (b) House 21; (c) House 22; (d) House 23. (a–c) gradiometry (top), resistivity (bottom); (d) gradiometry (top left), resistivity (top right), conductivity (bottom).

largely corroborated the remote sensing data; no hearths or internal pits were found; artifacts, such as there were, were sparse and consisted mostly of lithic expedient tools; and no datable material was found.

## Conclusion

Our survey data in some ways corroborate and in other ways challenge Rogers's (2011) characterization of Spiro's tripartite spatial division. The results indicate that the large-scale geophysical survey at Spiro is a powerful data set on its own merit. It becomes even more effective when combined with existing historical and archaeological data generated by both old and new excavations (Clements 1945; Rogers 1980, 1982a, 1982b, 2011; Rogers et al. 1989; Sievert and Rogers 2011) and the reanalysis of Craig Mound by Brown (1996, 2010, 2012, 2014). Perhaps more importantly, the deeper understanding of the Spiro landscape gained from the geophysical data provides novel and interesting ways of interpreting the site in relation to practices and ideologies in the Arkansas Valley, the Caddo culture area, and the broader Mississippian world.

Rogers's (2011) interpretation of the upper terrace platform and house mounds as special-purpose constructions for rituals related to death remains valid. In the Arkansas Valley, special-purpose submound structures tied to mortuary rituals have been documented at numerous additional sites via excavation and geophysics (Kay and Sabo 2006; S. Sullivan and McKinnon 2013), but data from off-mound contexts are lacking. In southeast Oklahoma, broad-scale geophysics has shown that the major Caddo mound centers were largely unoccupied (Hammerstedt et al. 2010; Regnier et al. 2014). This is unlike Caddo sites in Arkansas and Texas, which appear to have had resident populations living near mounds (see various articles in Perttula 2010; Perttula and Walker 2012). While Arkansas Valley sites including Harlan (Bell 1972) and Collins (S. Sullivan and McKinnon 2013) have roughly circular configurations of mounds, there are not enough archaeological data to determine whether these represent formal mound and plaza configurations common at Mississippian sites to the east. The previous interpretation of the central portion of the Spiro upper terrace as a plaza could be called into question, given the presence of a number of structure-sized magnetic anomalies on small, nearly imperceptible rises throughout this area. Testing these anomalies will help determine whether these were in fact buildings, whether they were built for residential or special-purpose use, and whether they were long-term features of the upper terrace architecture.

While they have not yet been ground-truthed, the linear pattern of anomalies at the interface between the upper and lower terraces is especially intriguing. They may represent an intentional segregation of space between the two mound precincts. This sort of spatial differentiation is common at Mississippian mound centers to the east on multiple spatial scales; notable examples include the palisaded central precinct at Cahokia (Fowler 1997) and the screened areas around the summit of Mound E at Moundville (Knight 2010:230–231). The central portion of the site, from Brown Mound east to Craig Mound, does appear to be a habitation area, with a series of structure-like anomalies and WPA-excavated houses extending in a rough arc several hundred meters north and west of Craig Mound. Although it may be a residential area, these presumed buildings are scattered and widely spaced, certainly not indicative of a robust residential population.

Finally, we turn to the lower terrace adjacent to Craig Mound, the portion of the site that Rogers (2011) characterized as related primarily to ritual disposal of the dead. For much of its use life, the area around Craig and Ward mounds was again primarily vacant, with the exception of a handful of houses excavated by the WPA. As the geophysical data have shown, for a brief time a series of short-term, haphazardly constructed buildings stood on the lower terrace. Unfortunately, at this time, we are unable to confirm a date for any of these buildings. When our new model of the Spiro landscape is combined with Brown's hypothesized Spirit Lodge ceremony, it is tempting to suggest that the geophysics has provided us a window into a transformative event at Spiro (following Beck et al. 2007; Bolender 2010), which drew in people from the surrounding region and fundamentally changed the Spiro social and political order. Without the geophysical survey, we would certainly never have located these ephemeral features and been able to reconstruct this important moment in Spiro's history.

## Acknowledgments

We thank the Tulsa District of the US Army Corps of Engineers (USACE), the Oklahoma Historical Society, the Caddo Nation of Oklahoma, and the Wichita and Affiliated Tribes for their support of this work. Funding and logistical support were provided by the Arkansas Archeological Survey, the Oklahoma Archeological Survey, and the University of Oklahoma College of Arts and Sciences. The USACE also provided funds for a radiocarbon date. We also thank Dennis Peterson, site manager of Spiro Mounds Archaeological Park, for his help and good cheer while in the field.

# 2
# Investigating Mississippian Community Organization with Geophysics

## Two Examples from the Tennessee River Valley

*Shawn M. Patch, Sarah Lowry, and Erin Pritchard*

This chapter explores the use of geophysics for broad-scale visualization of community patterning of two Mississippian sites in the Appalachian region of the Tennessee River Valley in northeast Alabama and east Tennessee (see Figure I.1). Geophysical surveys in this region are becoming more commonplace and have the potential to add significant new information to existing knowledge of Mississippian site organization and community patterning. With advocacy and sponsorship from the Tennessee Valley Authority (TVA), more large Mississippian sites will be available for study in the future.

## Research Design and Problems

As part of its responsibilities under Section 110 of the National Historic Preservation Act (NHPA) of 1966, the TVA is tasked with surveying, inventorying, planning, and managing archaeological resources under its control. In the past, compliance with this statute has typically been accomplished through shovel test surveys to locate additional sites. These efforts were not particularly useful for managing existing resources. Consequently, the TVA considered the application of noninvasive geophysical surveys in its planning and resource management program.

A pilot study was conducted at the Raccoon Creek Wildlife Management Area (WMA) in Guntersville Reservoir, Jackson County, Alabama (Patch and Lowry 2013). At that time, specific emphasis was placed on geophysics as a means of archaeological site identification, both to avoid inadvertent effects on sensitive features such as burials and to reduce the growing number of curated collections that result from traditional Section 110 excavations. In the course of this large survey, a highly detailed map of the Mississippian mound center known as the Cox site (1JA176) was produced.

Based on the success of the first study, a second investigation was conducted of the Bell site (40RE1), a large multiple-mound Mississippian town in Watts Bar Reservoir, Roane County, Tennessee (Patch and Lowry 2014). Refinements were made to the research design based on the results of the previous survey at the Cox site. One of the primary differences was the focus on greater resolution at a single site that was presumed to be more complex but about which very little was actually known.

The general objectives at the Cox and Bell sites were to (1) apply geophysical methods to prehistoric site identification; (2) identify intersite and intrasite feature patterning, site structure, and community layout; (3) compare the geophysical results with previously excavated similar sites in the same region; (4) assist with National Register of Historic Places (NRHP) eligibility recommendations and future nomination; and (5) produce recommendations for best management practices. An assessment of the initial research goals made it clear that the geophysical data from the Cox and Bell sites had tremendous potential to define site structure, feature patterning, and community organization at a scale that was inconceivable even a few decades ago.

## Mississippian Manifestations in the Tennessee River Valley

Mississippian settlements tended to be located on rich alluvial bottoms along major rivers and streams because these areas were particularly favorable for maize agriculture (Chapman 1985). Common site elements included earthen platform mounds with temples, elite residences, or council buildings, towns with individual households arranged around an open plaza, and palisades or ditch enclosures (Chapman 1985; R. Lewis and Stout 1998; T. Lewis and Kneberg 1946). House types tended to be square or rectangular, with individual posts set in wall trenches (earlier) or with single-post construction (later) (R. Lewis and Stout 1998; T. Lewis and Kneberg 1946; W. Webb 1938).

The regional chronology of Mississippian culture in east Tennessee has been delineated into four phases, commonly termed Mississippian I–IV (Pritchard 2009; Schroedl 2009; Schroedl et al. 1990). Mississippian I is a Woodland transitional period defined as the Martin Farm phase (A.D. 1000–1100) (Schroedl 2009; Schroedl et al. 1985, 1990; L. Sullivan 2016). The Mississippian II period is defined by the Hiwassee Island phase (A.D. 1000–1300) (Schroedl 2009; Schroedl et al. 1990). Dallas phase sites are placed within the Mississippian III period (A.D. 1300–1600) (T. Lewis and Lewis 1995; Polhemus 1987; Schroedl et al. 1990; L. Sullivan 2009). Finally, the protohistoric Overhill Cherokee sites make up the Mississippian IV period (A.D. 1600–1819) (Schroedl 2009).

Mississippian sites are well known in the Middle and Upper Tennessee

River Valley and many of its tributaries. This is largely the result of dam and reservoir construction and corresponding salvage excavations (T. Lewis and Kneberg 1946; T. Lewis and Lewis 1995; Polhemus 1987; W. Webb 1938, 1939; W. Webb and DeJarnette 1942; W. Webb and Wilder 1951). Many of the WPA excavations were constrained by time, funding, and the need to investigate as many sites as possible, often with the water literally rising as they worked. These factors resulted in a primary emphasis on mounds and comparatively limited excavations of associated village areas. One major exception to this pattern was W. Webb's (1952) work at the Jonathan Creek site in Kentucky, which had an explicit focus on defining the community pattern (Schroeder 2011, 2013). In subsequent decades, as the practice of archaeology matured, methods and research questions changed. Yet as major advances were made in method and theory, the number of large-scale excavations, with the exceptions of the Tellico and Townsend projects, dwindled primarily because no new major reservoirs were constructed. Despite extensive archaeological work in this region, several archaeological problems still exist related to the Mississippian period.

## Case Studies

The Cox site is a Mississippian single-mound town that was first investigated by Clarence B. Moore (1915). At the time of his visit, the site consisted of a flat-topped mound approximately 4 m (13 ft) high with basal dimensions of 27.4 x 31.4 m (90 x 103 ft). Archaeological investigations were limited to the village area because of a standing barn on top of the mound. Moore (1915) noted the presence of high artifact densities, extensive midden and shell deposits, and the recovery of more than 40 burials from the village.

Intensive excavations were carried out by the WPA prior to completion of Guntersville Reservoir by the TVA (W. Webb and Wilder 1951). By the 1930s, the mound was reported to be approximately 2.4 m (8 ft) high and 30.5 m (100 ft) in diameter, with a conical shape. Multiple floors and buildings were identified in the mound and approximately 109 burials were recovered. The conclusions reached by W. Webb and Wilder (1951) reflect the prevailing paradigm of the time, with an emphasis on basic cultural history. As indicated by the treatment of the mound and village as separate entities, there was no emphasis on overall site structure.

The Cox site today covers an area of approximately 10.8 ha. The boundary incorporates a much larger area than W. Webb and Wilder (1951) recognized because of a Section 110 survey (Gaffin et al. 2012). The site is currently in cultivation but is subject to no-till practices. Because of its location on the upper end of Guntersville Reservoir, it has not been inundated.

The Bell site is a multiple-mound Mississippian town in Roane County,

Tennessee. It first appeared in the literature in a reference by J. W. Powell (1884) and was the subject of a cursory visit by Clarence B. Moore (1915). Subsequent excavations were conducted by the WPA under Thomas M. N. Lewis in 1935 and Charles Nash in 1941 prior to completion of Watts Bar Reservoir by the TVA. Although the site was never reported formally, field notes and other documentation indicate the presence of six mounds, including one, Mound 51, that was approximately 9 m (30 ft) high (Patch and Lowry 2014). Excavations included the majority of Mound 51 and limited testing of Mounds 52, 53, and 54, but no work was done in the village area. Ahlman et al. (2000) conducted judgmental shovel testing of shell concentrations and made a surface collection of the eastern end of the site as part of an earlier Section 110 inventory. Results indicated extensive lithic debitage, Woodland and Mississippian ceramics and diagnostic projectile points, and intact features below the plow zone.

The Bell site today covers all of Huffine Island and measures approximately 8.1 ha. The island setting is a result of impoundment to create Watts Bar Reservoir, but the site is above the waterline. It is no longer farmed or plowed, and vegetation is restricted to grasses and early successional plants that are cut and cleared periodically by wildlife management staff. TVA stabilization efforts have focused on two of the mounds that are exposed to shoreline erosion and wave action (Fay 1987).

The Cox and Bell sites share many similarities, including strong Mississippian components, terrace settings along the Tennessee River, previous archaeological excavation by the WPA, comparatively little professional investigation since the 1930s, the presence of human remains, lack of information about existing archaeological features and their overall physical integrity, current locations as part of active wildlife management areas, and the need for better and more-informed management strategies and long-term stewardship. All of these characteristics make them ideal candidates for geophysical surveys.

## Archaeological Geophysics

Kvamme (2003a) promoted the idea that geophysical surveys can provide primary data for the study of cultural structures and features at the landscape scale (e.g., tens of hectares). Since that publication, other researchers have expanded the concept that geophysical data can provide unique and highly detailed perspectives on archaeological sites that go beyond simply identifying subsurface features (Aspinall et al. 2008; Conyers 2012; V. Thompson 2014; V. Thompson et al. 2011).

Geophysical methods are particularly applicable to larger sites and cultural landscapes, where traditional methods might provide only a small window into broader patterns. It is possible to map entire villages and surrounding landscapes; obtain information about site plans, structures, and layout; and study the relationships between individual site components such as house clusters, dumping grounds, public structures, storage and borrow pits, gardens, plazas, and fortifications (Gaffney et al. 2000; Kvamme 2003a; Toom and Kvamme 2002).

Despite a rich collection of archaeological resources and a long history of professional research, comparatively few geophysical surveys have been undertaken in the study region (Koerner et al. 2010; Patch and Gregory 2009, 2010; Patch and Lowry 2013, 2014; Patch et al. 2015). The case studies discussed here are among the first to apply geophysics to specific research problems related to the Mississippian period in this region, and the first to treat very large, multicomponent sites.

## Methods

Geophysical surveys of the Cox and Bell sites consisted of full coverage with magnetic gradiometry and selected coverage with ground-penetrating radar (GPR) (Patch and Lowry 2013, 2014). Instruments included a Bartington Grad601 fluxgate gradiometer and GSSI SIR-3000 GPR system with a 400 MHz antenna. Field data collection used standard grid sizes of 30 x 30 m. GPR data were collected over known features, such as mounds, and then in select areas after reviewing the magnetic data.

Magnetic data were processed using TerraSurveyor following the steps outlined by Kvamme (2006a). All data were destaggered to reduce variations in data collection pace, destriped to equalize differences between grids, and then clipped to remove especially high values that might obscure subtle prehistoric features.

GPR data were processed in Radan 7.0 following conventional steps that included setting time-zero and frequency filtering to remove background noise. Velocity analysis was conducted on data from each grid to estimate relative dielectric permitivity (RDP) values for each site prior to generating amplitude slice maps.

One of the major goals of a geophysical survey is to produce imagery that can be used to develop interpretations of potential cultural features, which is typically done through an anomaly classification system (Kvamme, Ernenwein, Hargrave, et al. 2006:234). For both the Cox and Bell sites, we relied on the previous excavation results, review of relevant regional literature, review of the potential range of feature types that might be expected, association of specific geophysical anomalies with archaeological feature types (if possible), quantification, and GIS data integration. GPR anomalies were classified based on their plan view morphology, size, relative amplitude, depth, and profile analysis. Gradiometer anomalies were classified by their nanotesla (nT) values. We applied the principle of correlations between certain feature types and their corresponding nT values to both case studies with the hope that statistical and spatial patterns might be detected (Kvamme, Ernenwein, Hargrave, et al. 2006; Pacheco et al. 2005, 2009a, 2009b).

The obvious noise from historical agricultural activities meant that the results had to be viewed in a critical manner and at best provided an estimate of the potential feature density and distribution. Extensive cultivation stripes are visible in the magnetic data at both sites. Many of the individual anomalies are likely the result of agricultural activities and could be objects such as tractor parts or implement fragments. Metal detecting and electromagnetic induction (EM) were not included as part of the original research design because of time and funding constraints, but these methods could help identify metal debris and therefore strengthen interpretations.

## Geophysical Results

### Cox Site

The Cox site showed a very high frequency of magnetic and GPR anomalies and clear patterning (Figures 2.1 and 2.2). The outline of the WPA excavation area from the 1930s is visible in the gradiometer composite, including four approach trenches and the large block excavation on top of the mound (see Figure 2.1). The approximate boundaries of the trenches were used to estimate mound dimensions of 28 x 20 m. There is no indication of the base of the mound at its interface with the original ground surface in either data set.

One plaza is present southwest of the mound and measures approximately 40 x 35 m. Compared with other areas of the site, it is relatively devoid of features, although it is surrounded by probable buildings.

One of the most noticeable features in the gradiometer composite map is a large ditch/palisade system (see Figure 2.1). With the river as one boundary, the ditch system is approximately 340 m long and forms a large *D* shape that encloses approximately 16,187 $m^2$ (1.6 ha). The size of the area enclosed by the ditch system is average for Mississippian sites in the region (Hally 2008; Hally and Kelly 1998; Schroedl 1998; L. Sullivan 1987). This ditch system is visibly distinct and wide and has few gaps, leading us to conclude it is largely intact.

A second ditch system was mapped northeast of the mound. Compared to the larger ditch system it is less distinct and appears to have been cut off.

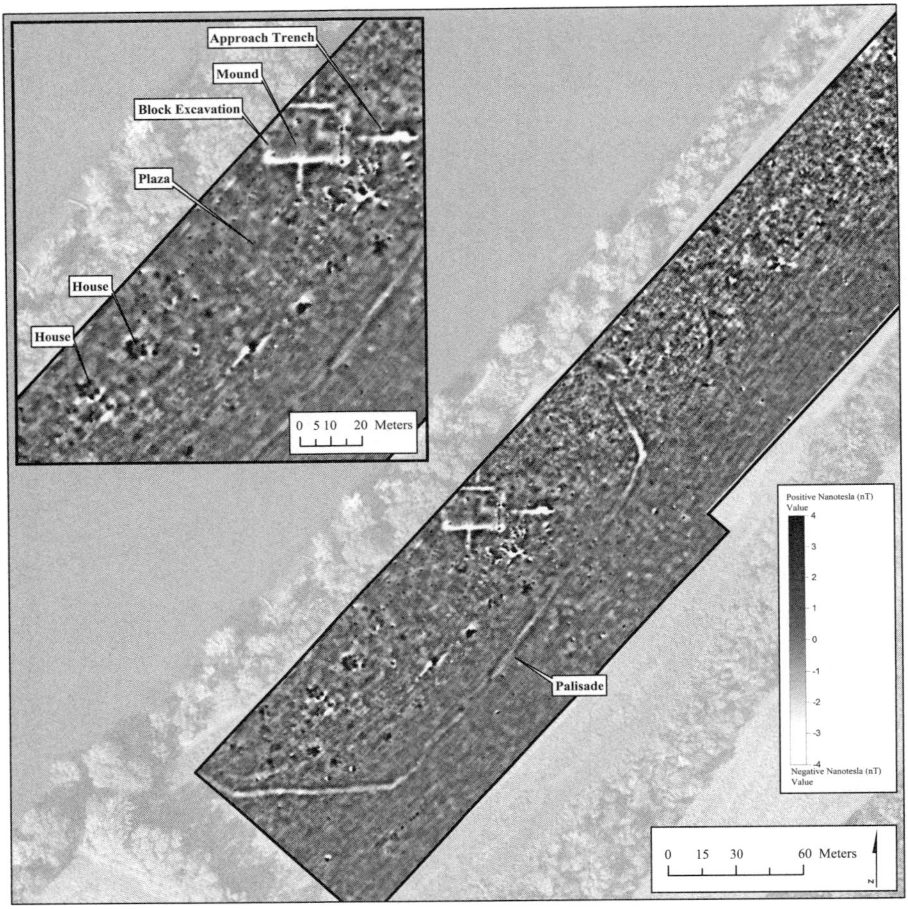


Figure 2.1. Gradiometer data mosaic of the Cox site (background from Microsoft Imagery, 2011).

These factors suggest that it was built at an earlier stage and was less developed and/or may be less preserved. The temporal relationship of the two ditch systems is unknown. It seems likely that the second ditch was an earlier extension of the larger ditch and that it was rebuilt to enclose a smaller space, which has been noted at other Mississippian sites (Polhemus 1987; Schroedl 1998). We hypothesize that these two ditch systems represent different building episodes.

There are several very clear examples of individual houses in the southwestern portion of the site with square or rectangular outlines (see Figure 2.1). The stronger magnetic signatures may be the result of burning or later building episodes in the site's history (Haley 2014a). Their shapes and sizes

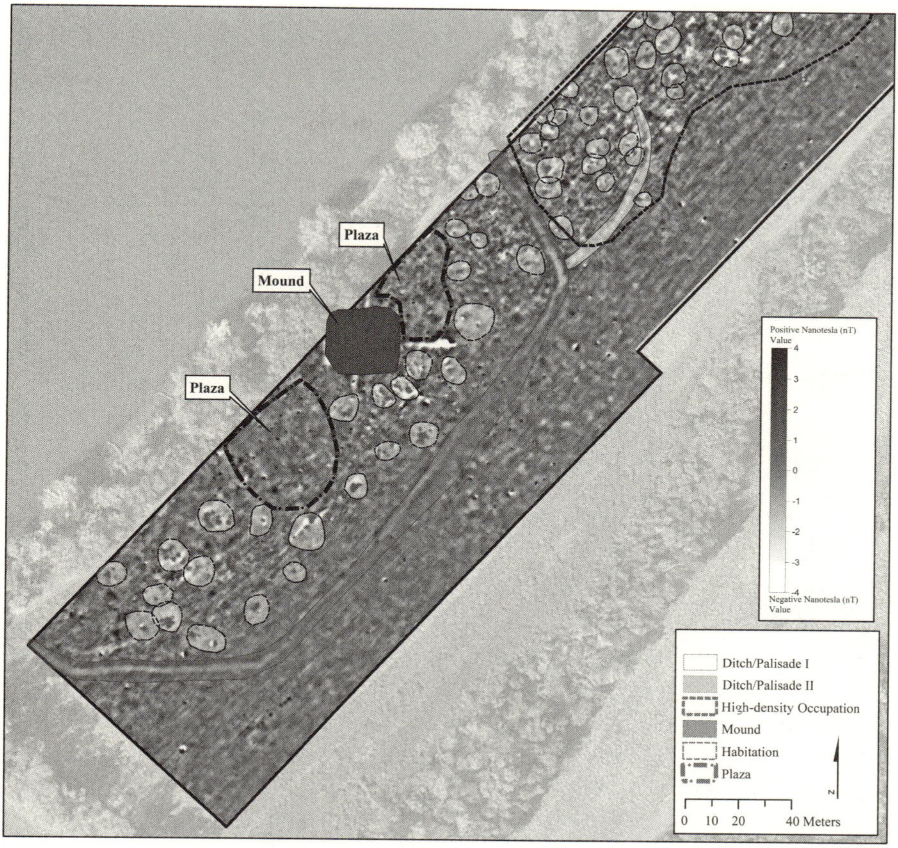


Figure 2.2. Interpretation of archaeological features at the Cox site (background from Microsoft Imagery, 2011).

are consistent with those of excavated houses at other Mississippian sites (Hally 2008; Hally and Kelly 1998; Patch et al. 2014; Schroedl 1998; Steere 2011; L. Sullivan 1995). Additional houses have been inferred from the patterning and clustering of magnetic and GPR anomalies. We recognize that these identifications are more tenuous because the physical contrasts are not as obvious.

The identification and spatial patterning of major site features suggest a well-planned community. The Cox site has many of the features found on other Mississippian sites: a single mound, a well-defined plaza, a ditch/palisade enclosure, buildings arranged around a plaza, and other features. However, it is one of the first in this region to be fully defined at the landscape scale.

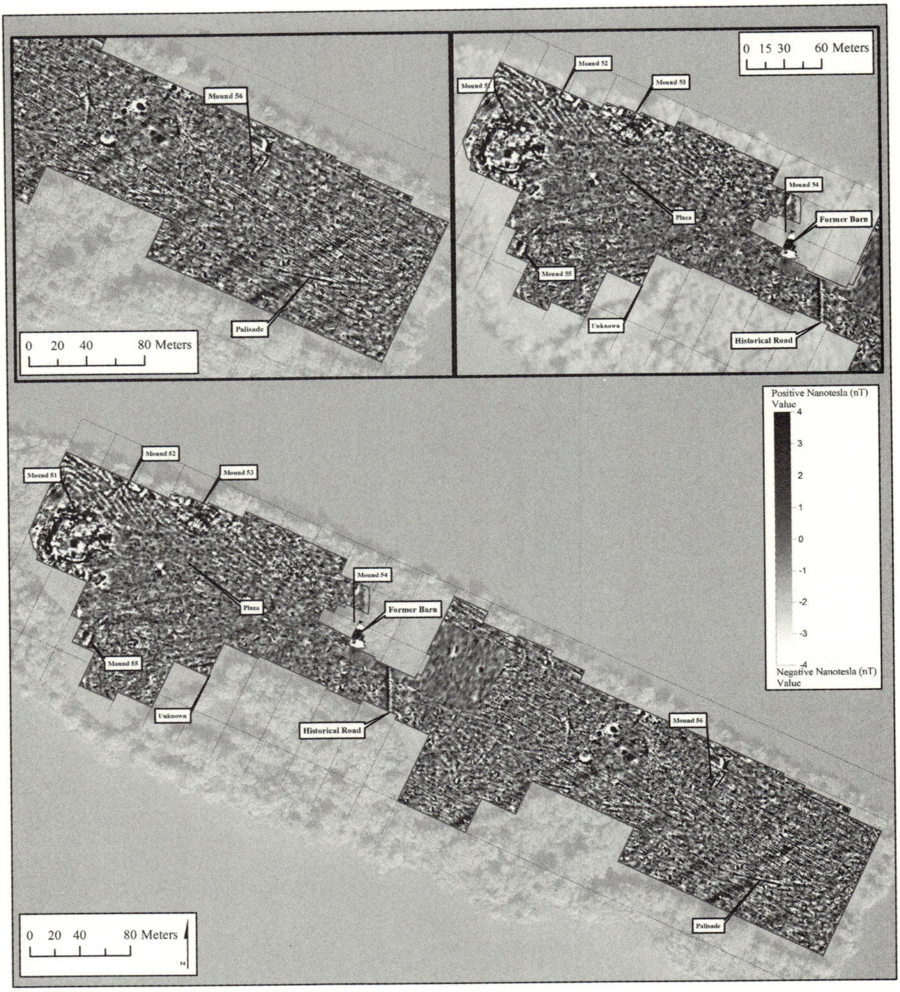


Figure 2.3. Gradiometer data mosaic of the Bell site (background from Microsoft Imagery, 2011).

## Bell Site

There is evidence in the geophysical data for extensive historical agricultural activities. Plowing is clearly evident in the form of cultivation stripes. Remnants of an old farm road that follows the tree line and runs parallel to Mound 54 are visible as a large magnetic anomaly. Previous research documented a barn on top of Mound 54, the remains of which are visible as a series of dipoles. It is highly likely that other historical metallic debris from farming is represented by many of the magnetic dipoles.

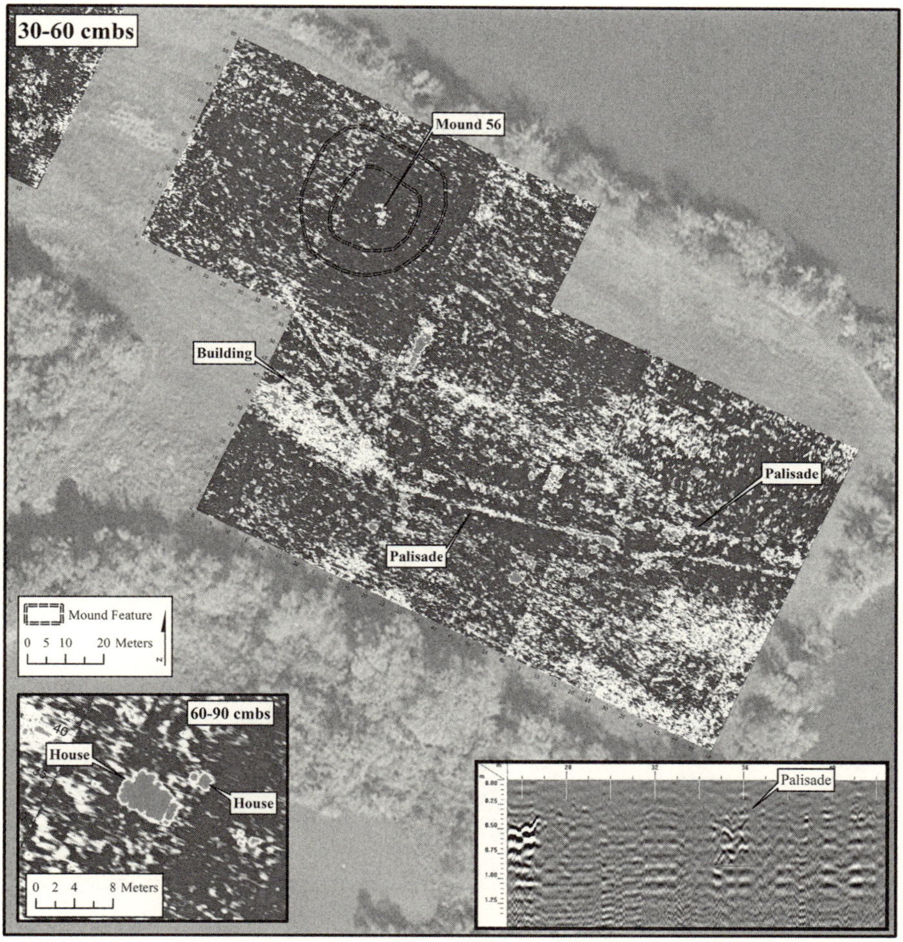


Figure 2.4. GPR slice map from 30 to 60 cmbs at the Bell site (background from Microsoft Imagery, 2011).

Despite the magnetic noise, geophysical data show major site features and provide excellent resolution of probable Mississippian community organization (Figures 2.3 and 2.4). At least three distinct spatial zones were identified (Figure 2.5). Zone 1 is the large mound complex and open plaza on the western edge of the site. Detailed excavation notes and documentation were found only for Mound 51, although other mounds were also investigated. Geophysical data sets for Mounds 51, 52, and 53 all show evidence of construction episodes and/or individual buildings.

The magnetic data for Mound 51 show a large, approximately square plat-

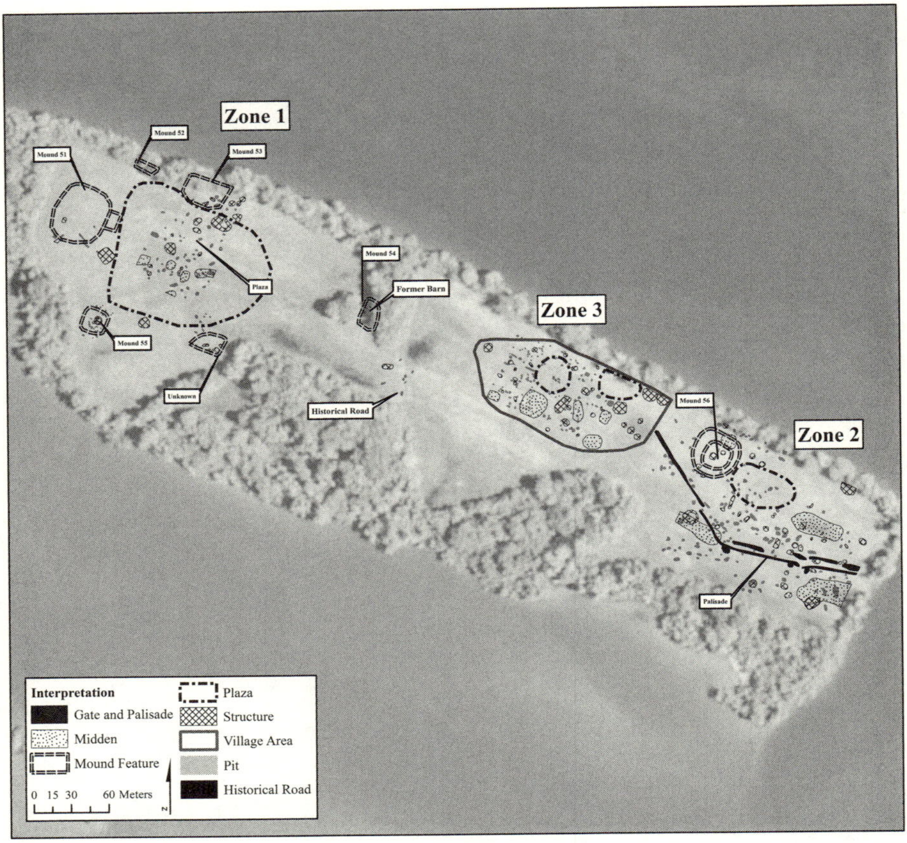


Figure 2.5. Interpretation of archaeological features at the Bell site (background from Microsoft Imagery, 2011).

form with a southeast-facing ramp (see Figure 2.3). Lewis's field notes suggest that this platform was observed during the excavations, and there was a specific question regarding whether or not it was burned. In this case, the magnetic anomaly is likely from clay that was either hard packed or burned toward the base of the mound, similar to the clay platforms at Hiwassee Island and Cemochechobee (T. Lewis and Kneberg 1946; Schnell et al. 1981).

Mound 52 is a platform mound with a small rectangular structure on its summit. Its boundary is incomplete, and an unknown portion of the mound has eroded into the Tennessee River. The rectangular structure appears to have been burned and was likely of wall trench construction based on its outline in the magnetic data. This mound is significantly smaller than

Mounds 51 and 53. The wall trench construction suggests that the structure on the summit dates to the Hiwassee Island phase.

It is not entirely clear whether Mound 53 was primarily a platform or burial mound, because there is evidence for both. Two burials were recovered, and the excavators inferred the presence of prepared floors even though no post molds were identified. The magnetic data for Mound 53 show the faint outline of a possible clay platform similar to that in Mound 51, but not as well defined. A clay platform likely indicates that the early portion of this mound was constructed in the Hiwassee Island phase and interments were made later in the same phase or in the beginning of the Dallas phase.

The geophysical results provide very little new information for Mound 54. The excavation notes indicate this was a platform mound with at least one structure, but there is no mention of construction methods or artifact data. Therefore, it is not possible at present to estimate the construction phase(s) of the mound. Dense vegetation on its summit and sides prevented an extensive geophysical survey. Magnetic data from the small area that was surveyed showed extensive interference from metal debris related to the barn documented here in the 1930s.

Geophysical data also defined two small mounds southeast of Mound 51. Today, these are both visible as minor rises above the surrounding terrain. Mound 55 (west) has a range of probable features but no obvious mound construction patterns. GPR maps of the mound to the east (not previously numbered) appear to have features that may be related to construction episodes and/or buildings. The origin, function, and purpose of these mounds are currently unknown, and several possible explanations might be considered: (1) they may be natural and not cultural, (2) they may be from the early Woodland period, or (3) they may be backdirt piles from the Mound 51 excavation. Currently, we have insufficient detail to resolve this issue any further.

All six mounds in this area surround what may be a large plaza that encloses an area of approximately 1 ha. As with the Cox site, overall anomaly density is significantly lower than in surrounding areas. In contrast to the Cox site, however, no buildings were identified immediately adjacent to the plaza. One explanation may be that buildings are present but do not have sufficient contrast to be imaged with either instrument. A second possibility is that buildings are farther from the main plaza and are now in wooded areas or submerged. A third explanation is the large volume of space occupied by the mounds themselves, which might have precluded dwellings close by.

Consideration of combined geophysical data and limited archaeological information for Zone 1 indicates both Hiwassee Island and Dallas phase occupations. There is evidence that Mound 51 was constructed primarily during the Hiwassee Island phase with minor modifications during the Dallas phase, and circumstantial evidence that significant portions of Mounds 52 and 53 also date to the Hiwassee Island phase (Patch and Lowry 2014). Absolute dates for early Mississippian platform mounds are becoming more common in east Tennessee (Koerner 2012; Koerner et al. 2010; Schroedl et al. 1990; L. Sullivan 2007, 2009). Thus, the potential Early Mississippian construction dates at the Bell site are consistent with those of other sites in the region.

Zone 2 was defined as the mound/palisade complex on the eastern end of the site. Mound 56 is the major surface feature in this area, measuring approximately 1,323 $m^2$ in area and less than 1 m in height. Both the mound outline and individual features are visible in the geophysical data sets. There is a faint outline of a probable square building on top, although it was not possible to determine the construction methods (e.g., wall trench or single post).

The nearly complete outlines of a prominent outer ditch/palisade system and a partial second inner system are visible in the GPR data (see Figure 2.5). The outer ditch/palisade minimally encloses an area that measures approximately 1.4 ha. The inner ditch/palisade is more segmented than the outer one and is present only along the southern boundary. It has a slightly segmented appearance that may indicate less physical integrity. The presence of these two systems suggests different construction or rebuilding episodes. The construction and rebuilding of palisades is a common feature at some Mississippian sites in the Tennessee River Valley (T. Lewis and Kneberg 1946; T. Lewis and Lewis 1995; Polhemus 1987).

The outer ditch/palisade has an unusual "angular" appearance with a distinct vertex. However, close examination of the Cox site ditch/palisade also shows at least two and possibly three distinct angles. Palisades at the Toqua site have multiple angles and were documented archaeologically to have completely surrounded the town (Polhemus 1987). The difference in appearance may be a result of how the palisade was arranged around the town. For example, a palisade that enclosed only three sides (e.g., Cox or King sites) might be more *D* shaped. In contrast, a palisade that completely surrounded a town may have been more angular. No historical roads or other features that were obviously associated with the mound were noted in this area, and a historical origin and function seems highly unlikely.

There are two palisade openings that likely represent gates or access points similar to those noted at the Dallas site (T. Lewis and Lewis 1995:335). In one instance, the palisade walls appear to overlap, which is consistent with palisades noted at other sites such as Hiwassee Island, Dallas, and Toqua (T. Lewis and Kneberg 1946; T. Lewis and Lewis 1995; Polhemus 1987).

A small plaza was identified east of the mound that measures approximately 1,313 $m^2$. As noted with the plazas in Zone 1 and at the Cox site, this one has fewer anomalies than the surrounding areas. Multiple inferred structures are distributed between the plaza and the palisade. They tend to have irregular outlines and are low contrast, which suggests they were not burned.

Several probable midden features are present in and around the palisade. Ahlman et al. (2000) noted the presence of a shell midden in the general vicinity. These are likely related to domestic activities associated with the village.

The types of features, their frequencies, and their spatial organization indicate a strong Mississippian component. The low platform mound, well-defined palisade, and compact nature of Zone 2 are consistent with other Dallas phase sites (T. Lewis and Lewis 1995). Therefore, we hypothesize that Zone 2 represents a Late Mississippian Dallas phase town.

Zone 3 was defined as a large area of high feature density between Zones 1 and 2 that contains numerous pits, middens, posts, and probable buildings representing dense village occupation. Many of the buildings are visible in the GPR results, with strong horizontal reflections that are interpreted as floors located below the prehistoric ground surface. Semisubterranean houses are characteristic of both Late Dallas and Mouse Creek phase sites such as Ledford Island (T. Lewis and Lewis 1995). Portions of this village appear to have cut into the western edge of the palisade in Zone 2, which suggests it postdates that occupation. The lack of a mound, the semisubterranean houses, somewhat random patterning, and presence of individual posts suggest that this portion of the site dates to the Late Dallas phase. Its relationship to Zone 1 is unclear; however, it is possible that some of the later occupation floors on Mound 51 were contemporary with the village deposits in Zone 3.

Defining and sorting the different temporal components at the Bell site is difficult, but not necessarily impossible, based solely on geophysical data with no published archaeological work for comparison. What we see in the geophysical data sets is a broad range of cultural features from multiple phases and occupations, all of which seem to overlap. The end result is a static image or snapshot of the site after its last occupation. Based on the limited archaeological information available, the regional Mississippian context information, and the spatial patterning in the geophysical data, we have offered tentative hypotheses for the chronological placement of Zones 1, 2, and 3. We anticipate that future work will refine these hypotheses.

The overall physical layout, with multiple mounds, a large plaza, and multiple buildings, indicates a substantial site. Many of these mounds, or at least mound summits, were probably occupied at the same time. The presence of multiple mounds from any time period around a single plaza is unusual in east Tennessee; the presence of multiple mounds that were occupied simultaneously may be exceedingly rare (Lynne P. Sullivan, personal communication 2014). Where and how the Bell site fits in the broader Mississippian world is a question that cannot be answered at present but should be given serious attention in the future.

## Discussion

We suggest that large-scale geophysical surveys represent a new and productive approach to Mississippian period research in the study area in the following ways. First, it is now possible to visualize the Cox and Bell sites at a landscape scale. Prior work at both sites was limited almost entirely to mounds, with little if any attention paid to the associated villages. Consequently, interpretations and conclusions derived from that work were limited and have not been revisited or revised in subsequent decades. New insights into spatial organization and site structure based on geophysical data provide a more complete view. We now know not only that both sites are far more complex than previously recognized, but that they share many of the common elements associated with Mississippian towns, such as plazas, ditch/palisade systems, individual houses, and feature clusters.

Second, the feature patterns were used to derive anthropological interpretations and hypotheses for further testing. For the Cox site, Patch et al. (2014) found that building sizes compared favorably to the size of buildings excavated at other contemporary sites. The calculation of area dimensions for plazas and ditch/palisade enclosures was possible for both the Cox and Bell sites. Comparison with other Mississippian sites in the region suggests significant variation in spatial organization among different towns.

Third, relative temporal changes can be inferred based on changes in the locations of ditches/palisades, houses, and overlapping features. The Cox site expanded and contracted through time and at least two ditch/palisade systems were built. The Bell site has three distinct zones that may indicate successive occupations from the Early through Late Mississippian period.

Fourth, site size and feature density can be used to infer the intensity of sociopolitical complexity. There is widespread recognition that Mississippian town sites varied considerably. Polhemus (1987:1240) suggested a five-level hierarchy of sociopolitical complexity for the Dallas phase that included Level I (households), Level II (household aggregates), Level III (towns), Level IV (town aggregates), and Level V (culture area or chiefdom). In contrast, L. Sullivan (2016), with more recent data, has suggested that Dallas and Mouse Creek phase sites are characterized by larger towns, not hierarchical settlement. Both perspectives are useful for evaluating the geophysical results.

Whether or not one accepts Polhemus's typology, the basic premise has merit. Levels I–III all have archaeological correlates at the site level, while Levels IV and V are abstract and consist of larger aggregates. In Polhemus's typology, both the Cox and Bell sites can be included in Level III (towns). However, the architectural features at each site clearly show differences that are likely due to a combination of social complexity and occupation length and intensity.

When the sites are compared to each other, the Cox site has a simpler community plan and the Bell site has a more complex community plan, despite sharing the same overall organizational structure. Archaeological and geophysical data suggest that the Cox site community structure developed during the late Mississippian period and that the site may have been occupied for a short time (Patch et al. 2014; W. Webb and Wilder 1951). Conversely, archaeological and geophysical data indicate that the Bell site occupation spanned the entire Mississippian period with several possible discrete zones, each with different temporal affiliations (Patch and Lowry 2014). In that sense, the Bell site is very similar to other sites such as Hixon, Hiwassee Island, and Citico, all of which are large and may have been unique (Patch et al. 2015).

One interesting question is what factors made the Bell site an attractive location for such a major occupation over a long period. If our occupation sequence for Zones 1 and 2 is correct, then Zone 1 may have been abandoned near the end of the Hiwassee Island phase. Otherwise, the establishment of Zone 2 in such close proximity to a much larger town does not make sense. Successive but distinct occupations suggest that there may have been a particularly important variable about that location.

Fifth, the geophysical results form the basis for future archaeological investigations at the Cox and Bell sites. With site-wide data now available, it is possible to develop an entirely new suite of research questions that are both site specific and more broadly applicable, such as construction methods and sequences for ditch/palisade systems, in-depth exploration of individual houses, plaza construction methods and use, potential correlations between magnetic anomalies within certain nT ranges and specific archaeological feature types, and mound construction sequences. Very limited archaeological investigations such as coring, shovel testing, or small-unit excavation can now be deployed with maximum effect and limited ground disturbance.

Additional feedback from these types of investigations will significantly enhance the value of geophysical data sets.

Finally, the comprehensive site mapping provided through geophysical surveys was not possible until very recently. Geophysical surveys of the Cox and Bell sites represent the first time since the Tellico project in the 1970s that the TVA has undertaken large-scale investigations of specific Mississippian sites. Much of what we know about the Tennessee River Valley comes from a relatively small number of sites that were excavated long ago. The political and financial situation today is completely different from what it was in the 1930s and 1970s: we can no longer conduct large-scale excavations. Even if sufficient time and money were allotted, there are concerns for Native American perspectives, curation challenges, and a genuine conservation ethic. Given these realities, geophysical surveys provide a viable and highly effective method for the investigation of complex Mississippian sites.

## Future Research

Large-scale geophysical surveys are a relatively new direction in the Tennessee River Valley, and the TVA is at the forefront of this approach. The application of such surveys to Mississippian sites has significant implications for addressing a range of research questions. For the TVA in particular, geophysical surveys have numerous benefits for planning and resource management. From an archaeological perspective, geophysical data sets significantly expand what can be learned about Mississippian sites.

We have attempted to show the validity of geophysical surveys as a productive means of investigation at Mississippian sites in the Tennessee River Valley. The case studies presented here are merely the first of many, and we fully expect to see improvements with future studies, not just in our own research but with others as well.

Additional field methods and equipment selection should be considered, such as electromagnetic induction, electrical resistance, electrical resistivity tomography, and possibly metal detecting. Although the data collection parameters are different and time and cost are always important, there are certainly benefits to each method. Most sites on TVA land, including Cox and Bell, have been in cultivation since the nineteenth century. The resulting interference from agricultural activities poses a challenge to geophysical surveys but can be addressed.

Geophysical surveys are a process and cannot end with a single map. Revisions and refinements should be expected. Archaeological excavations and more large-scale geophysical surveys of Mississippian sites in the region will provide new data for both intrasite and intersite comparisons.

## Acknowledgments

The Tennessee Valley Authority provided funding and logistical support for the Cox and Bell site surveys as part of its Section 110 management program. We thank Michaelyn Harle and Lynne Sullivan for their helpful comments and suggestions on an earlier draft and for responding to our many questions. Three anonymous reviewers provided constructive comments that helped us better organize our data and improve the interpretations. We appreciate their insight.

# 3
# Evaluating the Use of Community Space at Two Southeastern Mound Centers Using Magnetic Gradient and Surface Collection Data

*Duncan P. McKinnon and Bryan S. Haley*

The application of landscape-scale geophysical data collection allows for the visualization of buried cultural features and their relationship to site structure and community space. Frequently, these landscape data are examined and interpreted using a framework of theoretical characteristics and traits that define cultural phenomena, as documented in studies on architecture, material culture, and ethnography, where available (Bolton 1987; Griffith 1954; Hudson 1976; Kvamme 2003a; Lacquement 2007; Mehta 2007; V. Thompson 2014; Trubitt 2009). In this chapter, we present case studies from Battle Mound (3LA1) and Hollywood (22TU500), two Southeast mound centers where the distribution of architectural features and use of community space are evaluated using a landscape-scale geophysical approach (see Figure I.1).

In a broad sense, the two sites are part of a late prehistoric archaeological culture termed "Mississippian" (ca. A.D. 900–1700), largely defined by several characteristics, including a rich tradition of iconography, maize agriculture, earthen moundbuilding, and a developed settlement hierarchy (Blitz 2010; Cobb 2003; A. King and Meyers 2002; Lankford et al. 2011; Reilly and Garber 2007; B. Smith 1978, 1990). While both sites are considered Mississippian in the most general sense, occupants at Battle Mound were distinctively Caddo, situated on the western periphery of the Mississippian culture with a relatively independent evolution of cultural continuity and change (Perttula 1992; Perttula and Walker 2012; B. Smith 1990). When compared to Mississippian period groups throughout the broader Southeast characterized by a "sociopolitical organization that entailed hereditary ranking and centralized leadership" (Steponaitis 1986:388), the maintenance of distinctive social and religious institutions (Knight 1986), and differential status "clearly expressed in mortuary ritual" (Steponaitis 1986:389), the attributes that define Caddo culture (see Perttula 1992) are broadly similar to those of Mississippian groups situated east of the Caddo Homeland (B. Smith 1990). In the Yazoo Basin, the broad alluvial valley along the Mississippi River where Hollywood is located, professional attention has only recently begun to explore how the complex array of mound centers, villages, and other sites compares with better-known areas of the Mississippian world. For purposes of this chapter, a comparison of the use of community space at Battle Mound and Hollywood offers insights into the overall study of contemporaneous Southeast groups "as a means of more clearly understanding the development of Mississippian social forms and the Mississippian world as an interconnected [economic, social, and political] system" (A. King and Meyers 2002:115).

Within Southeast mound communities, evidence for settlement hierarchy is closely tied to the size, number, and spatial arrangement of earthen mounds and the distribution of domestic and community or ceremonial structures. Today, mounds of varying sizes are usually the most visible elements of these sites, although in many cases they have been adversely affected by either reduction or removal associated with historical and modern agricultural practices. While mounds are often still present on the landscape, understanding the distribution of off-mound architectural features is more elusive and traditionally requires large-scale excavations (see Collins 1997). Yet the moving of hundreds or thousands of cubic meters of dirt to expose large areas is typically not employed today for a variety of reasons (Knight 2007; Kvamme 2003a; Perttula et al. 2008). Thus, landscape-scale data and subsequent visualizations are rarely available in the Southeast without geophysical surveys. With this concern in mind, we discuss the use of landscape-scale geophysical surveys to define the distribution of off-mound architectural features that define the Battle Mound and Hollywood sites. To explore the variability of landscape use and cultural activities associated with different segments of the community, surface collections from each site are examined. The use of surface collection data demonstrates the importance of integrating geophysical with other types of archaeological data.

## Battle Mound

The Battle Mound site is a Middle and Late Caddo (ca. A.D. 1200–1680) mound center located along the Red River in Lafayette County, Arkansas. The site is centrally located within the Caddo Archaeological Area (Perttula 1992:8) and is about one mile east of the Red River channel where numerous channel scars, oxbow lakes, and back swamps characterize the landscape. Caddo people occupied the site over the course of several time periods, particularly after about A.D. 1200 (Hoffman 1970:163–164; Krieger 1949; McKinnon 2011, 2012, 2013a, 2013c; Perttula 1992:118; Perttula et al. 2009; Schambach 1982:4; Schambach et al. 1980; Valastro et al. 1972), although material collected from the surface demonstrates that the landscape use has a greater time depth (McKinnon 2017). The most prominent feature is a large multilevel platform mound—the largest in the Caddo area and one of the largest in the Southeast (Muller 1978; Perttula 1992:118; Schambach 1982:7). Two very low rises (0.5–1.0 m in height) are directly east of the mound and likely represent the plowed remains of the "four low humps and rises of the ground that long cultivation evidently had considerably spread," as documented by Clarence B. Moore during his visit to the site in 1911 (Moore 1912:566–567). Although these rises have not been tested, they are likely a combination of naturally occurring ridge and swale topography, where Caddo people utilized these well-drained natural levees as occupational areas (Hoffman 1970:810), and the deliberate addition of earth to create spaces reserved as low house mounds. Portions of the large mound were excavated in 1948 (Krieger 1949; McKinnon 2010a), off-mound surface collections were conducted in the early 1980s (Schambach et al. 1980), and a landscape-scale geophysical survey covering much of the site was conducted intermittently between 2007 and 2012 (McKinnon 2009, 2010b, 2013a).

## Surface Collection

The surface collection program was begun to document the spatial arrangement of artifacts in an initial attempt to define off-mound activity areas. Ten distinct collection areas were fairly broadly delineated based on the presence of topographic features, knowledge of previous archaeological finds (looter and avocational), and overall proximity and relationship to the large mound (Figure 3.1a). When artifacts were collected there was no "piece plotting" or accurate mapping of finds. As such, distributional information is confined or "clumped" to the scale of each collection area. Three areas are situated immediately at the mound base on the east (Area A), north (Area B), and west (Area C). Three areas are situated about 200 m east of the mound on low rises (Areas D, E, and J). About 100 m to the west are two areas (Areas H and I) on minor elevated landforms, and two other areas (Areas F and G) are situated 650 m northeast of the mound on low elevated natural landforms. All 10 surface collection areas show evidence of Caddo use and activity, with an early-twentieth-century non-Caddo historical farmstead located primarily within Area A (McKinnon 2017).

An analysis of the artifact types (ceramic, lithic, daub, bone), cultural typology, and associated densities demonstrates differential use in each area

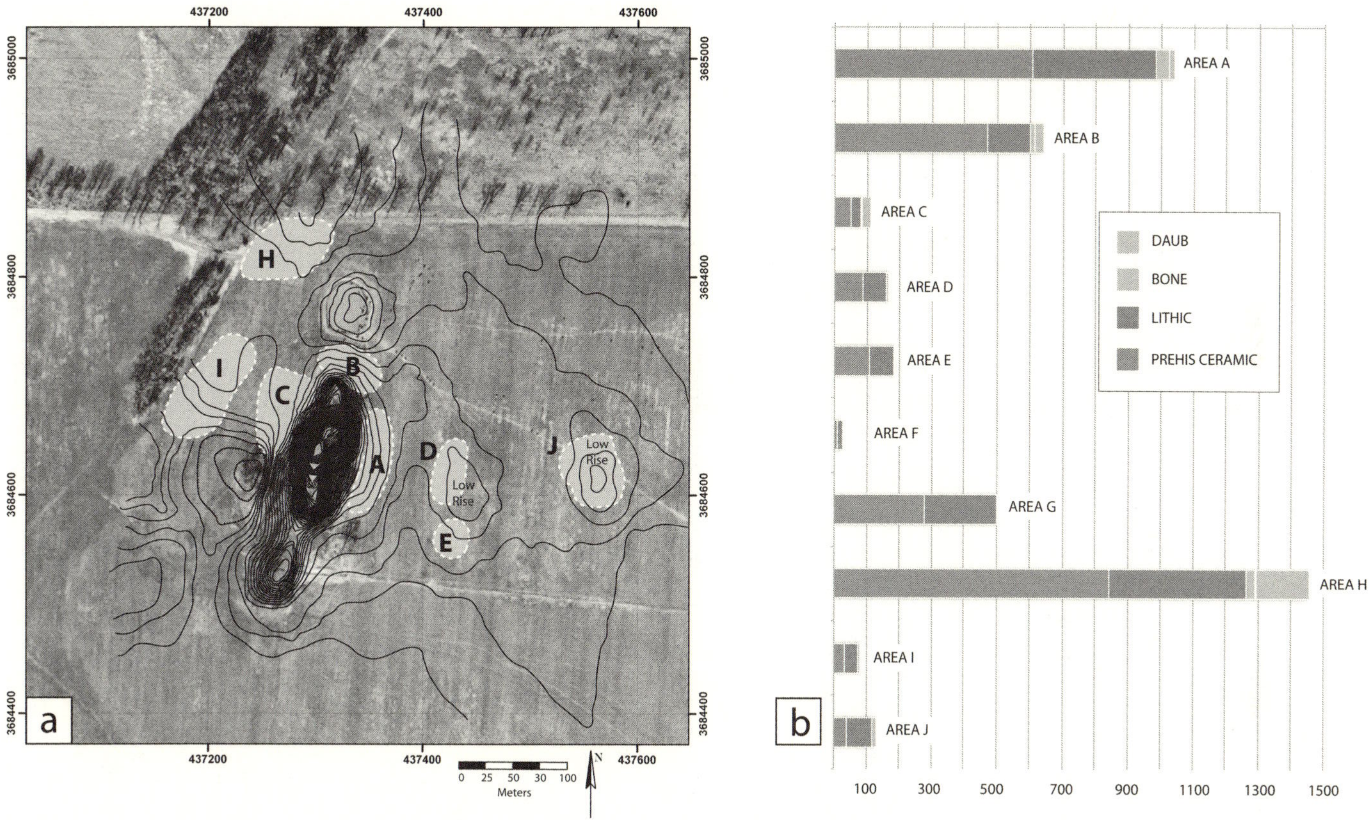


Figure 3.1. Map showing (a) surface collection areas at Battle Mound and (b) frequency of artifacts in each area. Note that Areas F and G are located 650 m to the northeast and are not illustrated.

(McKinnon 2017). Specifically, Areas D and E both contain low amounts of ceramic, lithic, and daub concentrations (see Figure 3.1b). Area J, farther east of the mound, also contains low amounts of ceramics and lithic debris, but higher concentrations (although minimal overall) of daub (see Figure 3.1b). When surface data are considered independently, the low number of artifacts in Areas D, E, and J suggests either the absence of significant long-term structures or the presence of minimal-use structures. Specific to Areas D and E, the low amounts of daub suggest the remains of structures that were either not burned or only minimally burned upon termination. Furthermore, the low number of ceramic artifacts in these three areas suggests they were not associated with long-term domesticity. Alternatively, the low number of artifacts might represent evidence of structures that were cleaned prior to the end of their use, reinforcing the idea of nondomestic use. As such, Areas D, E, and J, situated on low rises, are proposed to represent areas of activities tied to short-term communal rather than long-term domestic activities (McKinnon 2017).

Conversely, areas at the base of the mound to the east (A), north (B), and west (C) are proposed as domestic in composition and tied to daily events based on high amounts of utilitarian ceramic and lithic concentrations (McKinnon 2017). The higher concentration of surface artifacts in these areas (specifically A and B) suggests that daily activities (possibly domestic farmsteads; see Schambach 1982) are associated with the areas immediately proximate to the mound, whereas less domestic activities are reserved for areas on the low rises (D, E, and J) about 200 m east of the mound. Finally, Area H, situated about 100 m northwest of the mound, contained a significant number of artifacts and the highest concentration of daub. Area H is well known in the literature as the general location of the "Handy Place" cemetery. Many burials and hundreds of ceramic vessels have been excavated from this area, beginning with Moore in 1911 all the way through to the 1960s (Kitchens and Kitchens 1968; G. Martin 1939; Moore 1912; Perttula et al. 2009).

## Magnetic Gradient

Using the surface collection data as a spatial and interpretive framework, recent surveys were focused on gathering comparative, site-wide (14.32 ha) magnetic gradient data using a Bartington Grad601 gradiometer (Figure 3.2a). Results document numerous anomalies that represent characteristic shapes, sizes, and arrangements associated with prehistoric Caddo occupations. Cultural anomalies have been grouped as (1) those of historical origin; (2) those of Caddo culture origin; (3) those originating from natural processes; and (4) those of unknown origin (see McKinnon 2009, 2010b, 2017;

Walker and McKinnon 2012). The numerous anomalies document a complex Caddo settlement with differential use areas close to the large mound. Architectural features include several structures of various sizes, at least three community cemeteries, farmstead groups with compound fences, and borrow pits (Figure 3.2b) (McKinnon 2017).

The most abundant anomalies are circular patterns that range between 15 and 18 m in diameter for the largest and approximately 4 m in diameter for the smallest (Figure 3.3a, b). These are interpreted as Caddo circular structures, similar to those documented ethnographically (Bolton 1987; Griffith 1954:133), through excavation (Kelley 1997; Perttula 2005; Perttula and Rogers 2012; Schultz 2010; Trubowitz 1984; C. Webb 1940, 1959, 1983), and through geophysical methods (Creel et al. 2008; Perttula et al. 2008; Walker and McKinnon 2012; Walker and Perttula 2007, 2008). Importantly, at the Battle Mound site itself, three circular structures of similar dimension have been excavated from the mound platforms (L. Howard 1948; Krieger 1949; McKinnon 2010a, 2013a).

Several of the circular structures are situated within the three surface collection areas on low rises east of the mound (Areas D, E, and J). In particular, C28 and C29 (Area J) are similar in patterning to a structure excavated at the Late Caddo McLelland (16BO236) site, where several small- to medium-sized storage and trash pits with three burials formed a linear pattern aligning with the circle center (Kelley 1997:Figure 16). In addition, many anomalies have a concentric circle pattern (see Figure 3.3a, b), which may represent structures with concentric circles of wooden posts. A similar concentric circle pattern has been excavated at the Werner (16BO8) site, a contemporaneous Late Caddo mound site in northwestern Louisiana (C. Webb 1983:Figure 10). C. Webb (1983:217–221) interpreted the concentric pattern as the remnants of a large burned ceremonial lodge with an inner circle of roof supports and the larger outer circle comprising the structure wall. Within a standing structure, a concentric pattern of inner structure posts and an outer wall may have defined a space that was ritually cleaned.

An ethnographic example of a concentric circle structure with internal delineated activity areas comes from the Pawnee—a Caddoan-speaking group with traditional homelands north of the primary Caddo area (Carlson 1998; Hyde 1974). The example comes from structures associated with the Pawnee Purification of the Sacred Bundles ceremony (Linton 1923). The large Pawnee structure contains an inner circle of posts that delineate various ceremonial activities. As part of the ceremony, several priests (holders of sacred bundles) gathered in a lodge and seated themselves in "designated places around it, spreading their mats on the ground and hanging their bundles, unopened on the wall behind them [and] remained in the

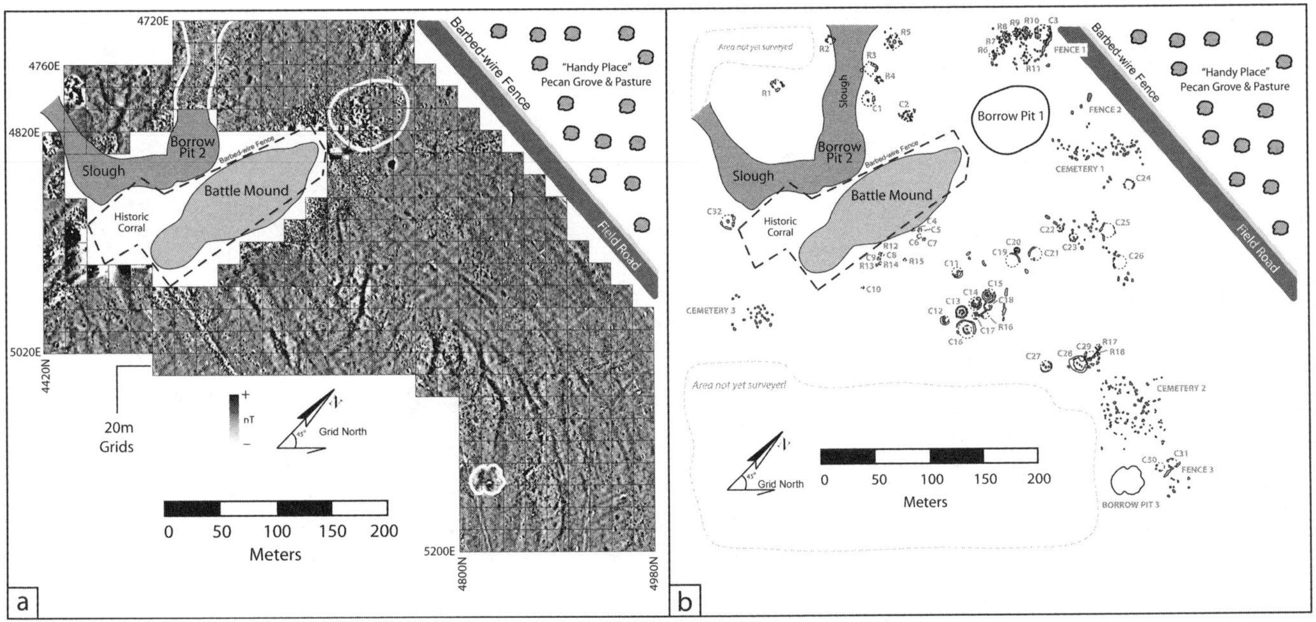


Figure 3.2. Results of (a) magnetic gradiometry survey at Battle Mound and (b) interpretation of results.

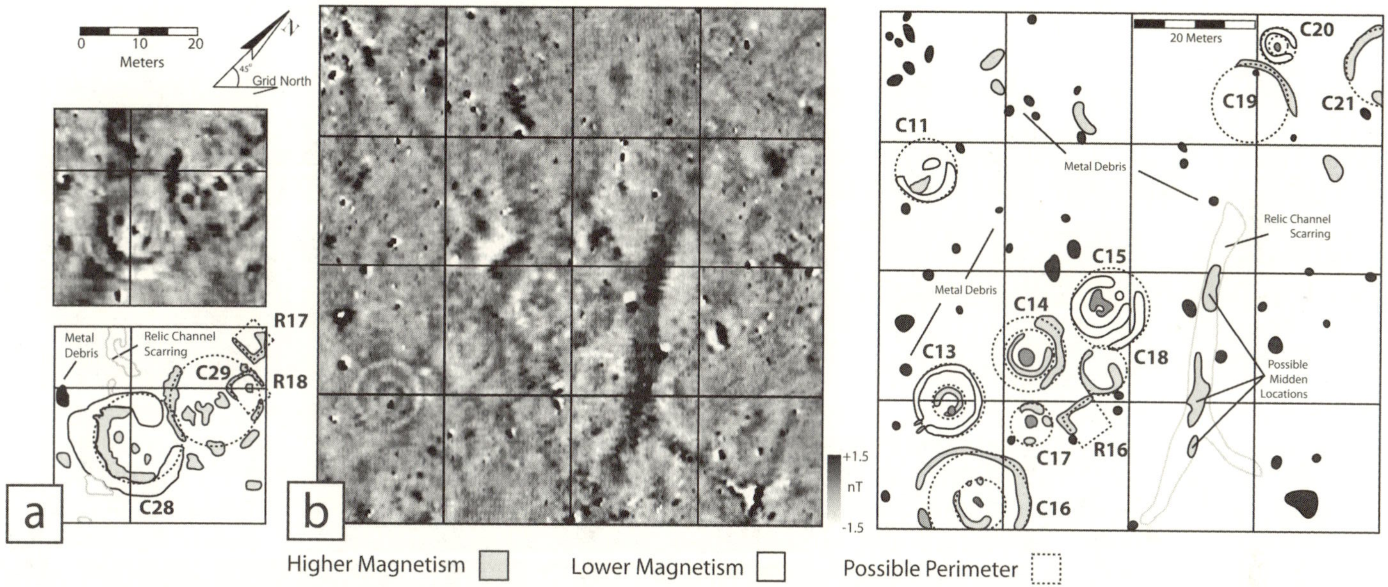


Figure 3.3. Examples of circular anomalies in (a) Area J (top and bottom boxes on left) and (b) Area D (two boxes on right). C, circular; R, rectangular.

lodge for three days and nights, sleeping in their designated places" (Linton 1923:41–42). Such activities could inadvertently remove magnetically enriched topsoil and leave the wide circular band of low magnetism seen in the circular anomalies at Battle Mound.

## Data Integration

The integration of surface material and geophysical anomalies offers insights into intrasite behavioral practices and associated differential use of community space associated with Caddo occupations at the Battle Mound site. Used together, geophysical and surface collection data document the remains of structures. Surface data document architectural artifacts, such as daub, and provide information on the type and quantity of artifacts in each area to supply a temporal framework. Geophysical data document several circular structures on the low rises. In other words, surface data suggest that structures might exist, whereas geophysical data demonstrate their presence, size, location, orientation, and arrangement.

The sandy low rises are a combination of ridge and swale topography and the deliberate addition of earth to create low house mounds. Such behavior would have been dually beneficial. Functionally, the permeable sandy soil provided a naturally drier area that was less prone to frequent flooding than the surrounding clay soil. Socially and symbolically, the low mounds represented an elevated space that befitted the elevated status of the inhabitant or inhabitants. The clustering of structures on two of the four documented low rises suggests social importance at the Battle Mound Caddo community. Furthermore, the difference in proposed architecture types found on the low mounds represents differences in function and social use (C. Webb 1940). For example, the large, high-magnetic walled structures on the low mound in Area J (see Figure 3.3a, C28 and C29) are proposed as meeting or council houses where it has been documented that the "council building was a large one, and the architecture apparently was different from that of the round grass lodge" (Bolton 1987:119). Relatedly, the low-magnetic cluster of several circular grass lodge structures in Areas D and E (see Figure 3.3b, C12–C17) is proposed as temporary or seasonal structures that housed "chief men" and "emissaries" from surrounding tribes during periods of ceremony and visitation (Bolton 1987:119; Griffith 1954:133).

As documented in ethnohistorical accounts, visiting relations between the Hasinai (see Bolton 1987) groups of east Texas and the Red River Kadohadacho (see Flecher 1907) were frequent and reciprocal. Neighboring groups would visit "en masse" after seasonal harvests to "renew their friendships, reaffirm their pledges of mutual aid against common enemies, exchange gifts, carry on trade, display prowess, and make merry together" (as cited in Griffith 1954:133). In preparation for these elaborate celebrations, individuals throughout the community were summoned together to provide food and construct lodging for important members of the visiting groups (Griffith 1954:133). If such temporary and palimpsest activities were reserved for the rise within Areas D and E, these events would explain (1) the lack of daub linked to the construction of temporary grass lodges (which would burn quickly and leave little thermoremanent signature [see Perttula and Skiles 2014]); (2) the low number of ceramics in these areas (since vessels would likely have been communally situated within the council house rather than in the temporary dwellings); (3) the close spatial relationship to the mound and council houses in Area J; and (4) the high concentration of low-magnetic circular structures throughout Areas D and E.

Using both surface collection material and geophysical results has offered insights into the use of space at the Battle Mound site during Caddo occupations, particularly after about A.D. 1200. When integrated intrasite behavioral practices and the use of space are summarized as a whole, communally based events are interpreted as centered on the eastern extent of the site, which included council houses and temporary dwellings for visiting dignitaries situated on the low rises about 200 m east of the mound (see Figure 3.2b).

## Hollywood

The Hollywood site is located in the northern part of the Yazoo Basin, an alluvial valley that stretches about 322 km (200 mi) south from the modern city of Memphis, Tennessee (see Figure I.1). This setting along the eastern bank of the Mississippi River was an ideal location for late prehistoric settlement, as attested by perhaps the highest density of Mississippian centers in the entire Southeast. Although there were better-known Mississippian political systems around Cahokia (see Pauketat and Emerson 1997), Moundville (see Knight 2010; Knight and Steponaitis 2006; Steponaitis 2009; G. D. Wilson 2008), Etowah (see A. King 2003), and other sites, the Yazoo Basin generally lacked large political centers. Instead, for reasons that are still being explored, there is a dense array of medium-sized mound centers, with Hollywood being one of the larger examples.

One of the earliest visits to the site by the professional community was by Calvin Brown, who described and sketched many mound sites around the state for his 1926 work *Archaeology of Mississippi* (1926:120–122). The other prominent early visit to Hollywood was by the Peabody Museum's Lower Mississippi Survey (LMS) team, who sketched the site and collected surface material (J. B. Griffin and Ford 1940; Phillips et al. 1951). Both of

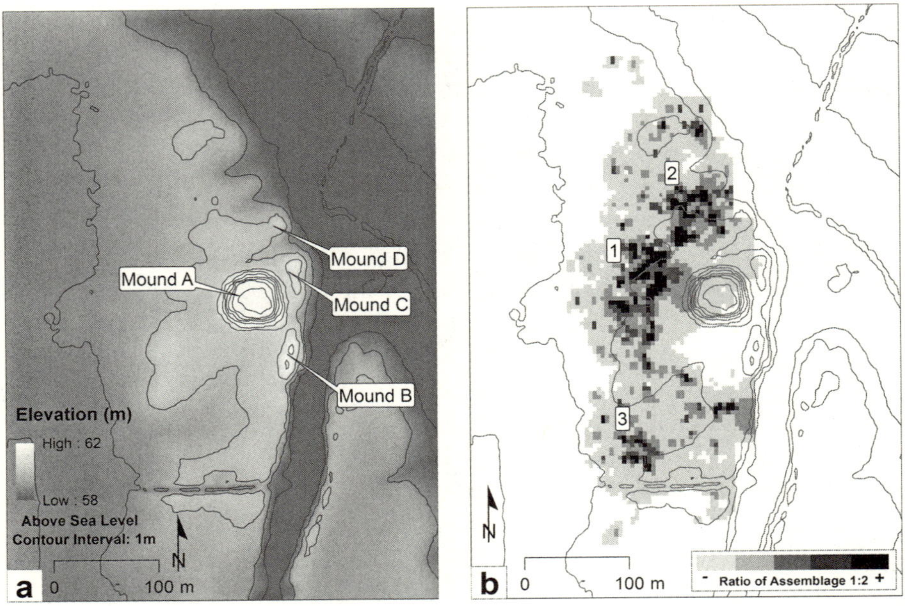


Figure 3.4. Map of (a) mound locations at the Hollywood site and (b) relative frequencies of Assemblage 1 to Assemblage 2 surface collection data.

these efforts provide information on the structure of earthen monuments at the site before it was greatly impacted by plowing. Both studies show the large central pyramidal mound and several smaller mounds extending along the edge of a prominent natural levee. Today, the largest mound (Mound A), which reaches a height of 7.9 m (26 ft), and three of the smaller mounds (Mounds B, C, and D) are still present (Figure 3.4a).

The Brown and LMS accounts differ in their depiction of the remaining earthen monuments, which are no longer visible topographically. C. Brown (1926: Fig. 24) illustrates a rectangular embankment that encloses the other mounds, while the LMS team (J. B. Griffin and Ford 1940) illustrates a collection of individual mounds. These two versions have implications for social and political structure at the site. Embankments and other types of barriers created limited access space for various ceremonial purposes and served as the boundary between public and private spheres (R. Lewis and Stout 1998:18–19). In contrast, a rectangular arrangement of mounds created a public space in the form of a plaza. In addition, in some cases (for example, Knight 1998), mounds organized around a plaza were closely tied to decision making, with clan leaders occupying structures on their summits.

Another important element of early research in the Yazoo Basin was the analysis of ceramics. As part of its seminal research, the LMS team analyzed surface-collected ceramics from a number of sites, including Hollywood, and established a chronology for the entire region (Phillips et al. 1951). Additional analysis of the Yazoo Basin collections by Phillips (1970) built on the earlier results and established a number of ceramic phases. Hollywood was included as one of the major mound centers of the Kent Phase (Phillips 1970:938–939), which we now know dates from roughly A.D. 1300 to 1500. Despite this early attention, as with most mound sites in the Yazoo Basin, little professional research was conducted there in subsequent decades.

## Surface Collection

Research interest returned to Hollywood in the late 1990s, when the University of Mississippi engaged in a multifaceted research program there, with one key element consisting of a systematic surface collection (J. Johnson et al. 2000:26–29). For that work, the entire site was divided into 2 m blocks and collections were made in a diagonal pattern of 5 percent of these, with 1,413 of the 1,575 squares producing artifacts. As with Battle Mound, the resulting data set is highly informative with respect to activities that took place across the site and their communal or domestic associations.

To understand intrasite behavioral patterns, the spatial distribution of ceramic types and varieties recovered during the surface collection was reconsidered for this chapter. The surface collection was divided into two distinct assemblages. Assemblage 1 consists of sherds that are slipped, engraved, or otherwise exceptionally well produced and generally corresponds to high status and mortuary contexts, and occasionally exhibits iconographic elements sometimes known as the Southern Cult (Phillips 1970:40–41, 58–59, 62–63, 142–143, 169–170). Assemblage 2 includes coarse ware sherds that are either plain or crudely decorated and often corresponds to utilitarian or domestic contexts (Phillips 1970:45–46, 130–135). A simple ratio was calculated for each block by dividing the number of sherds from Assemblage 1 by the number of sherds from Assemblage 2. A raster image was created from the ratio value using an Inverse Distance Weighted (IDW) algorithm in ESRI ArcMap 10.2 software.

In the resulting image (see Figure 3.4b), high relative frequencies of Assemblage 1 to Assemblage 2 are indicated by black, low frequencies by light gray, intermediate frequencies by darker shades of gray, and areas with no ceramics by white. Several spatial patterns are apparent from the resulting data. Two large and one small concentration of Assemblage 1 sherds can be identified, with the two large concentrations (see Figure 3.4b, 1 and 2) located to the west and north of Mound A and the smaller cluster (see Figure 3.4b, 3) visible to the southwest of Mound A. Note that while the summit of

Mound A was not adequately sampled during the surface collection because it was covered in vegetation, Assemblage 1 ceramics have been recovered in excavations there. Other areas exhibit lower ratios for the most part, meaning that utilitarian ceramics are prevalent, although there are small pockets of Assemblage 1 ceramics in other locations. Another pattern is the general lack of ceramics to the south of Mound A, which suggests this area could have served as a plaza.

## Magnetic Gradient

From 1997 to 2001, archaeologists from the University of Mississippi tested a wide range of geophysical and airborne remote sensing techniques, with the magnetic gradiometer emerging as the most informative (Haley 2002; J. Johnson et al. 2000; Peukert 2002; Reynolds 2002). Building on these initial surveys, a more detailed and larger-scale magnetic gradiometer survey was implemented during the summers of 2011, 2012, and 2013 using a Bartington Grad601 gradiometer (Haley 2014a). The resulting data (Figure 3.5a) yielded a wealth of information related to the spatial patterning of earthen monuments and structures and their relationship to community space.

One class of magnetic gradient anomalies present in the data from Hollywood is curvilinear and visible as arcs, ovals, or circles (see Figure 3.5). When this anomaly class forms near-complete ovals or circles, the maximum diameter is between 15 and 30 m. In several cases, the patterns are concentric, forming parallel arcs (Figure 3.6a). These anomalies are generally strong and dipolar, with positive segments ranging from approximately 10 to 30 nT and negative segments ranging from approximately -10 to -5 nT, although a close examination reveals several that are more subtle. Given the clayey soil in the Yazoo Basin area, the range of these anomalies suggests these targets were burned and exhibit thermoremanent magnetism (Kvamme 2006b:207–208, 216).

Several of these anomalies have been tested through excavation, revealing they are the result of burned daub deposited around the bases of mounds, before being sealed by mound construction episodes (J. Johnson 2005; Haley 2014a:45; J. Johnson et al. 2000:46). Based on the contexts of these features and historical accounts, this action was most likely an intentional means of disposing of burned house material (Haley 2014a:46–47). These mounds were later truncated and are no longer visible on the surface, but the sealed daub deposits remain. Once that identification was established, it became clear that these patterns closely match the individual mounds in the sketch map produced by the LMS team.

When the magnetic gradient data are coupled with a digital elevation model (DEM), the size and scale of extant and plowed-down mounds can

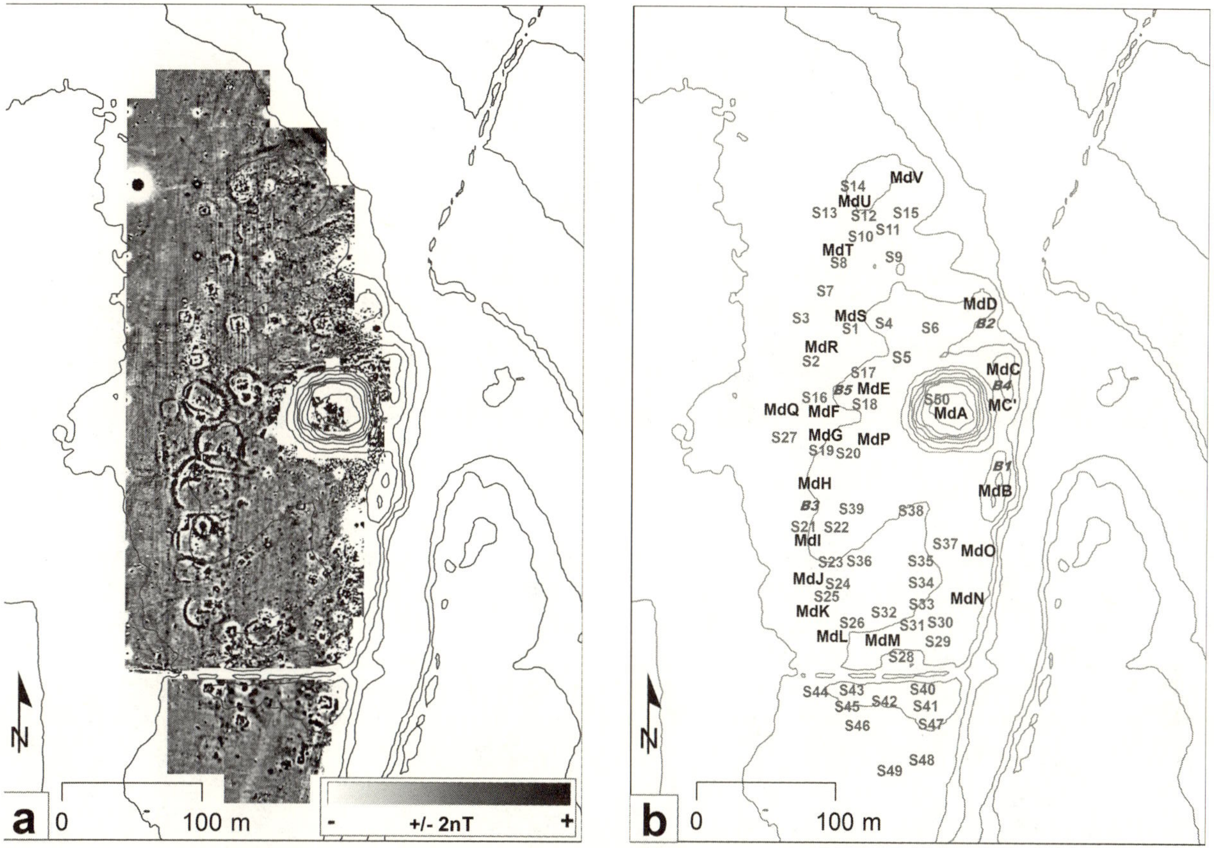


Figure 3.5. Results of (a) magnetic gradiometry survey at Hollywood and (b) interpretation of results. Md, mound; S, structure.

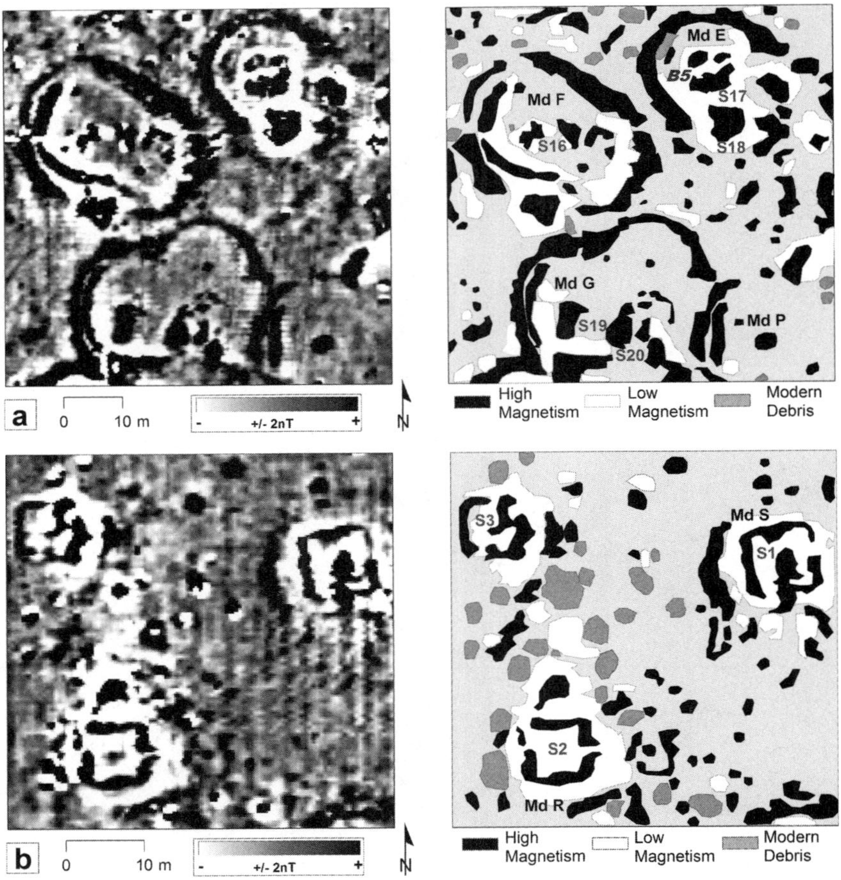


Figure 3.6. Select magnetic gradiometer results and interpretations highlighting circular anomalies (a) and rectangular anomalies (b). Md, mound; S, structure.

be assessed, providing some insight into political and social organization at the site. Estimates of volume for Mound A indicate a considerable investment in labor for its construction, which also implies coordination by centralized leadership (Milner 1998; Muller 1997). In addition to the social implications of its construction, this archetypal Mississippian platform mound was appropriate for a number of activities that served to reinforce social hierarchy (Hudson 1976:78; R. Lewis and Stout 1998:17; Lindauer and Blitz 1997:175). Excavations in the plaza to the south of Mound A indicate that it was artificially leveled using up to a meter of fill, which indicates that it too represents a considerable labor investment (J. Johnson et al. 2000:44).

The heights of mounds at Hollywood once ranged from 0.6 to 2.4 m

(2 to 8 ft), with the largest of these located adjacent to Mound A and gradually decreasing to the south (see Figure 3.5) (J. B. Griffin and Ford 1940). This may parallel a polarity trend identified at Moundville, where mound size, and the implied power and importance of the clans that occupied those mounds, also decreased from north to south (Knight 1998:50–51). It is not coincidental that enlargement episodes, marked by parallel arcs in the magnetic gradient data (see Figure 3.6a), are most prevalent to the north. J. Johnson (2005) suggests this is evidence for a general climate of competitive moundbuilding at Hollywood, where kin groups were motivated to outcompete their rivals to solidify their claim to power and status. Several smaller curvilinear anomalies, probably representing low house mounds, have also been identified to the north of Mound A (see Figure 3.5b). The extant Mound D, at 0.9 m (3 ft) tall, is one of these. If the mounds to the south of Mound A were the primary architectural core of the site, those that occupied and used the mounds to the north would have been excluded from the primary activity at the site.

The second class of features visible in the magnetic gradient data is partial square anomalies that range from 5.0 to 8.4 m on a side (see Figures 3.5 and 3.6b). Partial excavations of some of these have revealed the in situ remains of burned wall trench structures. Many of these are associated with mounds, which probably indicates that they were located on mound summits when they burned, although it is possible they are premound constructions that are still visible geophysically. In either case, their direct relationship to mounds implies that these were not typical domestic structures. The remaining rectangular anomalies are not clearly associated with mounds, although it is possible that some mounds lack the daub deposits that render them visible magnetically. Of the 50 total structures identified in Figure 3.5b, 15 are complete enough that the total area can be measured. These areas range from 25.5 to 70.6 m$^2$ (Haley 2014a:44–45), which, based on data from other Mississippian sites, suggests that most are elite residences or public structures (Scarry 1998:91–93).

In addition to their large size, the orientation and shape of the second class of anomalies are also informative. First, two structures (see Figure 3.5, S2 and S17) show evidence of extended central entranceways, which have been linked to mortuary ceremonialism in other parts of the Mississippian world (Perttula 2009; Trubitt 2009). In addition, one structure (see Figure 3.5, S1) exhibits an orientation of 7.8 degrees west of north, which is unique given the uniform cardinal orientation of all other structures. This anomaly is also conspicuous as the largest on the site, with an area of 70.6 m$^2$. Both lines of evidence indicate that this structure was particularly special and perhaps served as the residence of a powerful leader, the location of public or private ceremonies, or the venue for some other unique function. Note that this structure is located to the north of the primary mound architecture.

Considering all the magnetic gradient anomalies, it is noteworthy that such a large number are interpreted as various forms of burned structures. If the in situ burned structures and the secondary depositions around mound bases are included, there are roughly 100 and probably many more. Although unintentional, utilitarian, and militaristic explanations have been offered for burned structures at other sites in the Southeast (A. King 2003: 78–81; Scarry 1998:92; Trubitt 2009:242–244), excavation data from Hollywood indicate a widespread tradition of intentional burning. Considering ethnographic data—for example, the Tattooed Serpent procession witnessed by the French at the Grand Village site in 1725 (Le Page du Pratz 1975:337–339 [1774])—it is more likely that the burning of structures is ceremonial, perhaps commemorating the passing of elites or chiefly leaders (Hally 1996:95).

## Data Integration

When the surface collection and magnetic gradient data are compared, more can be said about community organization at Hollywood and activities that took place there. Two of the concentrations of Assemblage 1 ceramics located west and southwest of Mound A (see Figure 3.4b, 1, 3) correspond to the mounds south of Mound A (see Figure 3.5b). This association further supports the hypothesis that elites or clan leaders occupied these mounds, as they would presumably have had better access to these status goods, and that the summits of these mounds were used for ceremonial activities.

Alternatively, the prevalence of Assemblage 1 ceramics to the north of Mound A is unexpected. This area appears to contain only small house mounds, which are isolated from the proposed architectural core. Nevertheless, the area to the north of Mound A also contains burned structures that are unique in size, shape, or orientation. In addition, a number of burials have been found in this area (Stallings 1994), and some of these ceramics may have been burial goods or may have been associated with mortuary ceremonialism. Another possible explanation is that this region is not contemporaneous with the southern half of the site, although at this point that is not supported by ceramic frequencies or AMS dating (Haley 2014b).

While the plaza is generally clear of ceramics, the presence of small patches of ceramics belonging to Assemblage 1 within the plaza may indicate that this area was also used for ceremonial purposes, although in a more public or communal context. Based on the magnetic gradient data, there was at least one burned structure within the plaza (see Figure 3.5b, S38), although it is not as large or as intensely burned as others. This structure presumably played a role in the activity here and was perhaps associated with the Assemblage 1 material culture found here.

Perhaps of most importance, the presence of Assemblage 1 ceramics across much of the site indicates that ceremonial activities were an important site function and that a majority of residents were high status, part of elite lineage or kinship, or at least had access to prestige goods. This contrasts with surrounding villages and smaller mound sites where these fine ceramics are much less common. Therefore, Hollywood most likely served as a political capital and a periodic destination for lesser elites and other residents of the polity. The activities that they engaged in, probably initiated and administered by the high-status individuals residing at Hollywood, would have served to reinforce social and political hierarchies across the polity.

## Conclusion

As presented in this chapter, Battle Mound and Hollywood are examples of contrasting forms of Mississippian community organization. While many details differ, key Mississippian elements are still presented—most notably large, platform, earthen mounds, which are closely tied to social, political, and religious spheres (Knight 1986; R. Lewis and Stout 1998). Earthen moundbuilding alone is evidence that both these sites were capitals of their respective polities in the centuries before contact. As such, both sites functioned as the political, economic, and ideological focal point of the surrounding region.

Battle Mound, as a significant Caddo mound center, was spatially diverse and segregated, with some areas serving as the setting for short-term communal activities, others serving a mortuary function, and still others containing primarily domestic occupations (McKinnon 2017). While the large mound was clearly the focus of the site, the combination of surface material and geophysical survey data demonstrates that the function of activities within the community space was also important within overall site organization and integration into regional politics and social structure. Historical accounts of Caddo groups suggest that this function was to host ceremonies that strengthened ties with allied groups.

In contrast with the data from Battle Mound, geophysical data from Hollywood paint the picture of a more traditional mound center. The key structural elements include a number of mounds of varying sizes, a plaza defined by those mounds, and mound-top residential structures that were probably occupied by various kin groups and that facilitated some role in decision making for the site and surrounding polity (R. Lewis and Stout 1998). What is more unusual is the prevalence of burned structures on those mounds, on smaller house mounds, and off mounds. One interpretation (Haley 2014a) is that the burning of structures represents the commemoration of elite or special-status individuals upon their death.

The analyses presented in this chapter indicate that community organization at Battle Mound and Hollywood was decidedly Southeastern, with the structural configuration reflecting and reinforcing marked social hierarchy (Steponaitis 1986). At Battle Mound, this took the form of a large mound, smaller rises that supported council houses and temporary dwellings for visiting dignitaries, and a residential population that was obligated to support the communal activities that took place within those structures. At Hollywood, the configuration of mounds symbolized and reinforced social and political organization, and, if burned structures marked commemorative events, those activities enhanced the power and social rank of those kin groups, even those that did not reside on the primary mounds.

In closing, the conclusions presented in this chapter would be nearly impossible to arrive at without landscape-scale geophysical data. Alternatively, the geophysical data alone would be much less useful without integration with other archaeological data, in this case systematic surface collections. Indeed, geophysical data are most successful when integrated into a larger research design (see V. Thompson et al. 2011). When coupled with appropriate ethnographic data, such as the Pawnee at Battle Mound or the Natchez at Hollywood, these integrated data can offer great insight into the structure and operation of community space in the Mississippian world and beyond.

## Acknowledgments

Duncan McKinnon would like to thank Ken Kvamme, George Sabo, Jami Lockhart, Jamie Brandon, the University of Arkansas, the Arkansas Archeological Survey, and the numerous students and volunteers who helped with geophysical data collection. Bryan Haley would like to thank Jay Johnson, John Connaway, the University of Mississippi Department of Anthropology, the University of Mississippi Archaeological Field School, the Mississippi Department of Archives and History, Christopher Rodning, the Tulane University Department of Anthropology, the Mississippi Archaeological Association, the Tunica Museum, and Richard and Brooks Taylor.

# 4
# At the Tip of an Amplitude Wave

## The Role of Terrestrial Remote Sensing in Twenty-First Century Grand Canyon Archaeology Debates

*Philip B. Mink II*

Preeminent southwestern archaeologist Emil Haury was a firm believer that the best place for archaeologists to settle disagreements was at the tip of a shovel. He reasoned that the thorough collection of well-documented excavation data was the best method for two opposing sides to argue points of contention (R. Thompson 2004). However, in the twenty-first century, excavation is increasingly limited by a variety of factors including decreased funding, federal and state cultural resource regulations, and increased sensitivity to indigenous peoples. Terrestrial remote sensing provides twenty-first-century Southwestern archaeologists with new methods, sympathetic to these contemporary challenges, for obtaining data to test hypotheses and settle debates.

The focus of this chapter is to demonstrate how terrestrial remote sensing methods can contribute to modern Southwestern archaeological field work. When geophysical surveys are used in conjunction with smaller-scale excavations and the region's rich archaeological and ethnographic records, the old view that geophysical surveys have little value in the Southwest, with its prolific surface archaeological record, can finally be put to rest. Using examples from research into the Pueblo period (A.D. 600–1300) archaeology of the Grand Canyon, my research demonstrates how geophysical surveys can answer questions about archaeological site organization and prehistoric settlement practice.

### Geophysics in Southwestern Archaeology

Terrestrial remote sensing investigations of archaeological sites in the Southwest were first employed inside national parks (Johnston 1965; Maruca 1980; Vickers and Dolphin 1975). The earliest example of the use of terrestrial remote sensing in Southwestern archaeology is Richard Johnston's 1961 magnetometer survey of a buried structure at Mesa Verde National Park

(Johnston 1965). Data from that survey were initially plotted by hand and were unsuccessful at locating any subsurface features. However, when data were replotted a decade later on some of the earliest National Park Service (NPS) computers, a variety of buried archaeological features were identified (Maruca 1980). While no other geophysical surveys of Southwestern archaeological sites were published in the 1960s, the next decade saw an increase in the application of these techniques to the archaeology of the region.

For example, Chaco Canyon was the site of several types of terrestrial remote sensing projects in the 1970s. The Chaco Center, a joint research endeavor between the NPS and the University of New Mexico, conducted magnetometer (Loose and Lyons 1976), ground-penetrating radar (Vickers and Dolphin 1975), and electrical resistance (Vickers et al. 1976) surveys. All the surveys were successful at locating features associated with the prehistoric occupation of the site and supported continued research into employing geophysical surveys at Southwestern archaeological sites (Mathien 2005; see also chapters 5 and 7 this volume).

Beginning in the 1980s, terrestrial remote sensing began to be applied throughout the Southwest to undertakings required by the National Historic Preservation Act of 1966 (Public Law 89–665; 16 U.S.C. 470 et seq.), including both Section 106 and Section 110 investigations, a trend that continues today. Terrestrial remote sensing investigations are used either to guide excavations in a cost-effective manner (Bevan 1983; DeVore 2002; Hackbarth 2012) or to support federal land management agencies in the administration of their cultural resources (Abbot et al. 1997; DeVore 1996; W. Johnson 2003; Mink 2013). Because Section 106 is one of the largest drivers of archaeological research today, an increase in the use of terrestrial remote sensing for eligibility determinations is warranted. From the Section 110 perspective, geophysical surveys provide federal land managers with data on subsurface cultural resources without excavation. These data are useful not only for the management of resources (e.g., establishing boundaries, identifying buried features) but also for research on the archaeological sites under their purview that often cannot be excavated because of legislative or budgetary constraints.

In addition to compliance research, increasing numbers of academic investigations employ terrestrial remote sensing techniques in Southwestern archaeology (Conyers 2012; Conyers and Cameron 1998; Ernenwein 2006, 2008; Hinz et. al. 2008; Sternberg 1987; A. Sullivan et. al. 2012), albeit in a much smaller number of cases than in the compliance arena. These studies tend not only to highlight the archaeological applications but also to contribute to the larger field of near-surface geophysics.

The frequency of noncompliance projects employing terrestrial remote

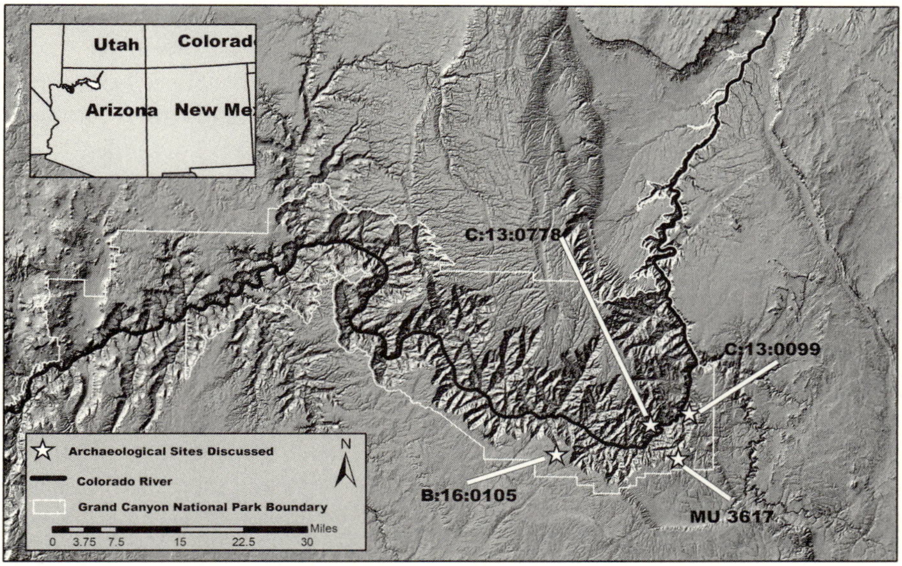


Figure 4.1. Approximate locations for sites discussed in this chapter.

sensing in the Southwest, while increasing, is still much lower than in other parts of North America. This muted utilization of geophysical techniques is the result of several factors including rugged natural environment, lack of access to equipment, absence of training at the major degree-granting institutions in the region, the "black-box" nature of the technology, and belief that these techniques do not provide much beyond what can be discerned from the surface archaeological assemblage. As I will demonstrate, the rugged nature of the terrain can be overcome and geophysical surveys can contribute to a fuller interpretation of the Southwestern archaeological record beyond the surface data, which should encourage increasing utilization of these techniques.

## Grand Canyon Case Studies

In order to consider the role of terrestrial remote sensing in twenty-first-century Southwestern archaeology, case studies from my research at Grand Canyon National Park are presented (Figure 4.1; see also Figure I.1). The case studies demonstrate the utility of geophysical surveys in answering questions about prehistoric Pueblo period (A.D. 600–1300) settlement at Grand Canyon and also show how these techniques can help researchers and land managers overcome the challenges of practicing archaeology in the twenty-first century.

Grand Canyon National Park is located in northern Arizona along the southern and western edge of the Colorado Plateau. It is over 446 km (277 mi) long and begins at Lees Ferry, 24 km (15 mi) downriver from the Glen Canyon Dam and Lake Powell, and ends beneath the waters of Lake Mead at the Grand Wash Cliffs near the Arizona and Nevada border. The canyon is between 1,067 and 1,829 m (3,500–6,000 ft) deep and the rims are between 1.6 and 24.2 km (1–15 mi) apart.

Native peoples have inhabited the canyon for at least 10,000 years (Fairley 2003; Mink 2015). The most intensive occupation occurred from A.D. 1050 to 1150, during the Formative or Pueblo II period (Mink 2015; Schwartz 1980). During this time, three distinct archaeological traditions have been identified within the present-day boundaries of Grand Canyon National Park, including two Ancestral Puebloan branches, the Kayenta and the Virgin, and a more local group, the Cohonina, whose settlement was centered on the San Francisco Peaks near present-day Williams, Arizona (Fairley 2003; Mink 2015; Schwartz 2008).

The presence of these three distinct archaeological traditions within the ecologically diverse Grand Canyon provides an excellent laboratory for understanding prehistoric cultural variation related to internal site structure and settlement patterns. My research focuses on variability in settlement practice from A.D. 600 to 1300 as an approach for investigating prehistoric adaptation. Terrestrial remote sensing investigations have provided information on subsurface archaeological deposits beyond traditional surface survey data on internal site structure that can be used to make improved inferences about prehistoric settlement.

Beginning in 2004 I began conducting limited terrestrial remote sensing surveys on previously recorded archaeological sites within Grand Canyon National Park. The techniques utilized were originally selected based on equipment availability, but as more instruments were acquired by the University of Kentucky, the available tool set increased and the techniques employed were determined by research questions and the conditions at the site. To date I have employed ground-penetrating radar (GPR), fluxgate magnetometry, and electromagnetic conductivity to over a dozen sites in the park. In the case studies presented below, two primary terrestrial remote sensing techniques were applied: GPR and magnetic gradiometry.

A fluxgate gradiometer, Geoscan Research's FM256, was employed to conduct the magnetic surveys. For these case studies, readings were collected at the 0.01 nT level every 0.25 cm along transects spaced 50 cm apart. Data were downloaded and processed in Geoscan Research's Geoplot 3.0 software package and then exported for visualization in Golden Software's Surfer 10 mapping program.

The GPR instrument used to collect data for these case studies was a Mala, Ramac GPR CU II Geo System with a 500 MHz shielded antenna. The GPR unit rested on a plastic skid plate and was attached to a distance-measurement wheel or a four-wheeled cart that was pulled or pushed by the operator back and forth across the grid on 50 cm transects. The 50 cm transects were utilized to ensure sufficient and total subsurface coverage. The data collected during the field survey were initially processed in the field and then more intensively processed at the University of Kentucky William S. Webb Museum of Anthropology in GPR Slice V7.

## C:13:0778 (UN-52)

Site C:13:0778 is part of the Unkar Delta site complex located on a large debris flow delta on the Colorado River at mile 72.5. The site was partially excavated by Douglas Schwartz (1980) and identified at that time as site UN-52. Schwartz described the site as containing two groups of structures and a third isolated room. Testing of the site revealed that the surface features were underlain by a midden that extended at least 20 m outside the structures. The chronology of the site is complex. A $^{14}C$ date of A.D. 950 ± 90 was obtained from one of the completely excavated rooms in Structure 1. However, the surface ceramic assemblage indicates both an earlier Pueblo I occupation (A.D. 900) and a later occupation of A.D. 1100–1180 (Pueblo II) (Schwartz 1980).

A portion of the site was overlaid with a 20 x 20 m grid, and magnetic gradiometer data were collected across the entire grid. However, because of concerns about damaging vegetation and cryptobiotic crust, GPR data were collected on only the eastern half in a smaller, 11 x 20 m section of the grid. The 20 x 20 m grid contains a portion of the midden area, all of Structure 2, which was partially excavated by Schwartz (1980), and possibly a segment of the fully excavated Structure 1. The smaller GPR survey covers the entirety of Structure 2 and extends from the terrace edge back 11 m.

The GPR results (Figure 4.2) indicate that the area examined has been heavily impacted by modern behavior, likely by both the 1960s excavations and the modern hiking trail through the site. The consistent sheets of high-intensity reflections and the lack of reflective material in three-dimensional rendering are caused by the distribution of screened archaeological fill and disturbances from foot traffic across the site. Slightly deeper, the GPR results indicate the tops of the walls of the partially excavated Structure 2, where both external walls and internal disturbances can be discerned. The disturbances inside the room blocks were likely caused when the excavations were backfilled with wall fall.

The most intriguing results from the GPR survey indicate a possible un-

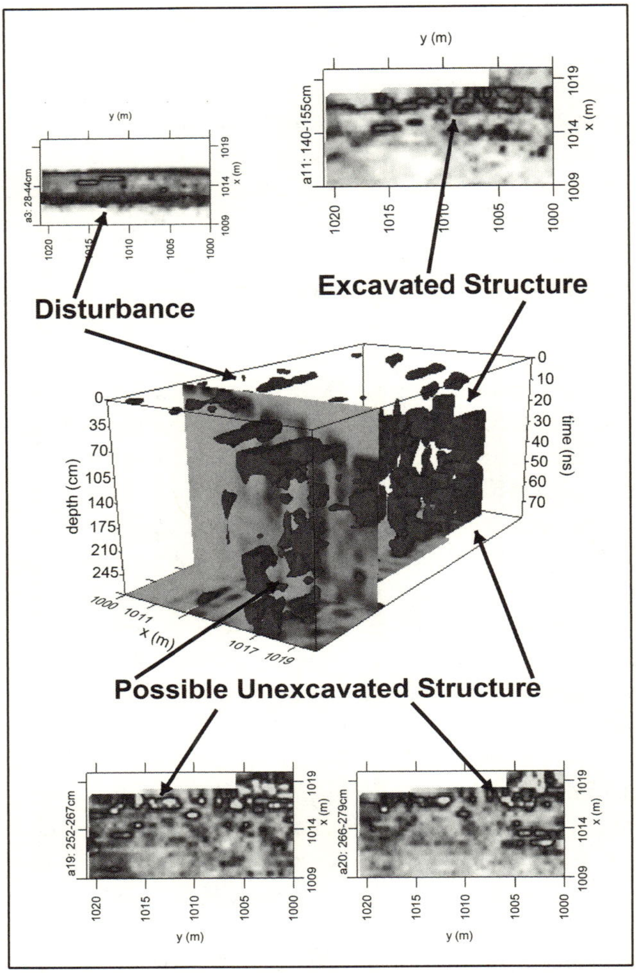


Figure 4.2. Results of ground-penetrating radar survey at C:13:0778 (UN-52).

excavated structure below the previous investigations. These walls expanded Structure 2 south with the addition of a couple more rooms. A structure (or two) with several rooms is easily discernible in slices a19 and a20 (see bottom of Figure 4.2), and, in the three-dimensional rendering of the data, both sets of results indicate a complex internal site structure that presents an unexpected view of local settlement practice.

Recent research by Anderson and Neff (2011) seems to indicate that the Pueblo period farmers in the Grand Canyon relocated their sites as they gained experience living along the often-dynamic Colorado River. They posit a shifting settlement strategy through time, as occupants were exposed to large floods. The GPR evidence from C:13:0778 indicates two occupation episodes at two distinct times in this locale on Unkar Delta and challenges that hypothesis. Their premise may be true along the more entrenched portions of the eastern Colorado River in Grand Canyon, but these GPR results suggest that the shifting settlement strategy was not universal.

Furthermore, numerous short-term settlement models have been proposed for the Pueblo periods at the Grand Canyon (Mink 2015). Many of these models posit a settlement system that entails seasonal mobility for exploitation of a variety of subsistence resources (Fairley 2003). Two settlement models are based on local ethnographic information (Fairley 2003; McCoy 1990). The Havasupai model suggests summer occupation in the canyon bottoms, and fall and winter occupation in the South Rim uplands (J. Martin 1985). The Paiute model hypothesizes residential bases in the North Rim uplands and springtime resource exploitation in the canyon bottom (Fairley 2003). Three additional settlement models are archaeologically based, including the Powell Plateau model (Effland et. al. 1981), which speculates that Pueblo peoples moved seasonally between garden houses and large pueblos in a single canyon environmental zone (North Rim, South Rim, or Inner Canyon). The Cross Canyon model posits hunting and gathering of pinyon nuts and other wild resources on the South Rim in the fall and winter, gathering agave and other wild plant resources along with hunting and fishing in the Inner Canyon during the spring, and farming in the Inner Canyon bottom and on the North Rim during the summer (A. Sullivan et al. 2012). The final model, the Biseasonal Settlement model (Schwartz 1980), posits a less short-term strategy with permanent residence in the canyon bottom supplemented by summer farming on the North Rim.

The GPR data from C:13:0778 suggest a more permanent settlement strategy where a site was used either seasonally or permanently during a variety of time periods. These data seem to support both Schwartz's Biseasonal Settlement model and the Cross Canyon model.

The magnetic gradiometer data (Figure 4.3) from C:13:0778 indicate a very large (~ 8 x 8 m), diffuse, circular anomaly in the central portion of the grid. It could represent a natural depression or a disturbance caused by inadvertent modern recreational behaviors, or it could be the remnants of the midden documented by Schwartz (1980).

In addition to the circular anomaly, a variety of other anomalies are related to the archaeological record of the site. Approximately 14 m north of Structure 2 is a series of anomalies where the fully excavated Structure 1 is

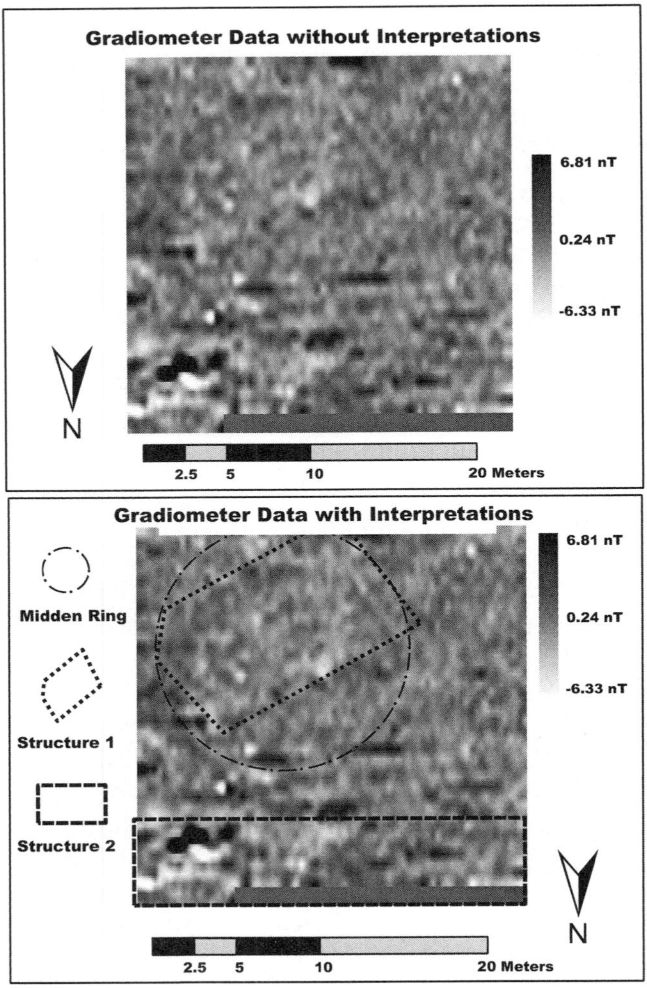


Figure 4.3. Results of magnetic gradiometer survey at C:13:0778 (UN-52).

located. The complete excavation and backfilling of the site left little surface evidence, and the compacted floors are the likely cause of the weak magnetic anomalies. While the anomalies identified as features are not incontrovertible, they are intriguing and are located where the records indicate there should be archaeological remains. Also, inside the large circular anomaly are smaller monopolar anomalies that could be cultural features such as trash or storage pits. There are also several elongated, monopolar, positive magnetic anomalies adjacent to the walls of the partially excavated Structure 2. While they could be prehistoric features, it is likely that they are test units associated with the earlier excavations. Inside the boundaries of Structure 2 are several positive monopolar magnetic anomalies and one large dipolar anomaly. All these anomalies can be correlated with similar anomalies in the GPR data and are likely associated with the backfilled archaeological features.

The geophysical survey at C:13:0778 was productive, as it resulted in the identification of some buried features that are prehistoric and others that are related to the 1960s excavations. For example, where the survey grids overlap, both techniques identified the remains of the earlier excavations. Additionally, below the partially excavated Structure 2, the GPR data identified several possible prehistoric anomalies that are unobservable in the gradiometer data set because of its limited depth. Within the area of the site that was surveyed with only the magnetic gradiometer (because of vegetation), a large circular anomaly was identified that likely represents the prehistoric midden originally identified during earlier excavations. Evidence of the fully excavated Structure 1 was also identified in the gradiometer data, within the larger circular anomaly, along with a variety of smaller monopolar anomalies that are possible pits or storage features.

The geophysical results obtained from C:13:0778 highlight the importance of employing multiple geophysical techniques (see Clay 2006). Variation in site soils and the archaeological record can affect the ability of a specific technique to locate the diversity of archaeological phenomena that may be present. In addition, survey conditions, such as heavy vegetation, or the depth of the deposits may restrict the use of a particular technique. Thus, the use of multiple techniques increases the likelihood of identifying cultural features.

The geophysical surveys at C:13:0778 not only identified buried features that will help NPS better manage this resource but also furthered our understanding of prehistoric settlement in the canyon. The GPR results of Structure 2 identified a previously unknown structure beneath the surface structure that was tested by Schwartz (1980). This earlier structure demonstrates that the site had a complex history with multiple occupations through time—a fact that was also indicated by the earlier excavations of Structure 1, which also contained evidence of reuse. These factors suggest that the shifting settlement patterns previously proposed for Colorado River sites may not apply to the entire river corridor.

## MU 3617

MU 3617 is an amorphous scatter of prehistoric ceramics, lithics, and daub fragments dispersed around a linear rock alignment (Figure 4.4). At the time of documentation, there was disagreement among the crew on how to record the site. Some argued that the rock alignment and associated ar-

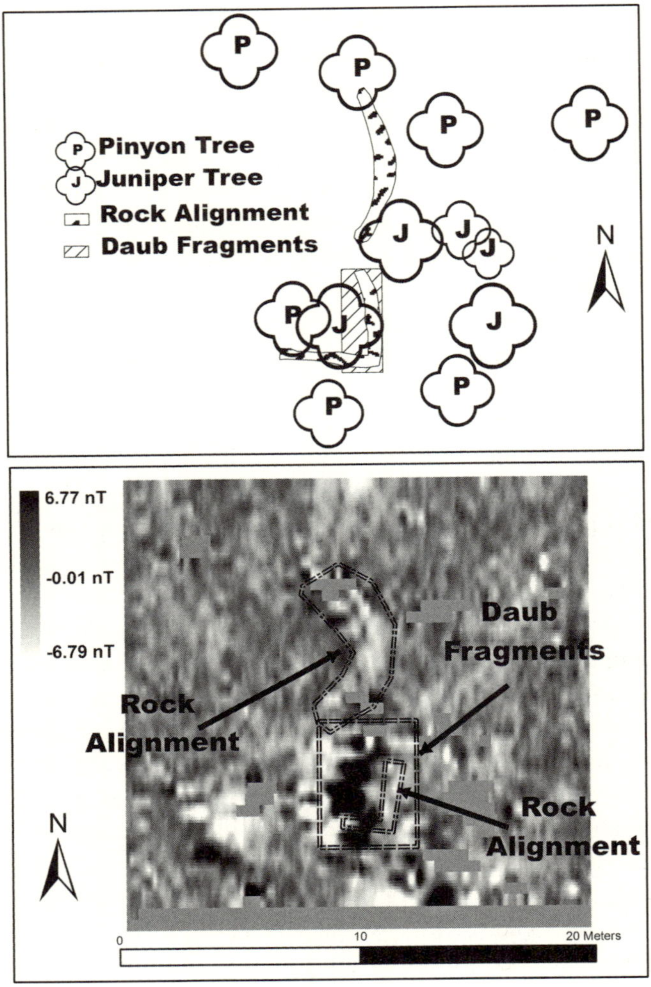


Figure 4.4. Results of magnetic gradiometer survey at MU 3617.

tifacts were evidence of a structure, albeit with portions of the wall buried or removed (a situation observed at other sites nearby). Others argued that the site was just an artifact scatter with a linear stone feature that could be related to a variety of nonstructural functions. To help settle this debate and properly record the site type, a magnetic gradiometer survey was conducted.

Results from the survey settled the debate. The wall is visible, but no evidence of a buried structure is present in the data. Moreover, the gradiometer survey provided a surprise: a large nebulous dipolar anomaly located south of the linear rock feature. Closer field examinations of the anomaly location revealed a large amount of daub and other evidence of burning hidden at the base of two pinyon trees and beneath the pine needles. These factors and the ceramic artifacts suggest that the site is likely the remnant of a prehistoric jacal or ramada structure. These temporary structures would have been inhabited by a small family group for maybe a year or less. The identification of this jacal structure provides additional evidence of a mobile settlement strategy along the South Rim of the canyon (A. Sullivan et al. 2002). When these data are combined with the GPR evidence of more robust and permanent structures in the Inner Canyon at site C:13:0778, it appears that Schwartz's Biseasonal Settlement model, which hypothesizes permanent settlements below the canyon rim that were supplemented with temporary settlements on the North Rim, can be expanded to include the South Rim, as proposed by the Cross Canyon model.

## Terrestrial Remote Sensing Signature Database

While the inferences presented above are based on solid evidence, they could be improved if the source of an anomaly could be positively identified. One of the biggest challenges of conducting terrestrial remote sensing surveys on lands managed by the NPS is confirming the origin of an anomaly through excavation. A variety of federal regulations and sensitivity to present-day Native peoples limit excavations; subsequently, determining the source of an anomaly is often challenging. One approach is to develop a long-term plan to conduct geophysical surveys on all sites prior to their excavation as part of the federal compliance process. These data can then be utilized to develop an anomaly signature database that can be later applied to sites that are not excavated, akin to the spectral signature databases used for satellite remote sensing studies (A. Sullivan et al. 2012). Two additional sites investigated with terrestrial remote sensing techniques in Grand Canyon National Park (C:13:0099 and B:16:0105) provide an excellent start to such a signature database. I have recommended that Grand Canyon National Park proactively develop such a database as part of its management strategy for the park's cultural resources and have begun compiling that data for all my geophysical surveys in the park (Mink 2013).

### C:13:0099

Site C:13:0099 is part of a site complex located on an extensive debris flow and delta on the left bank of the Colorado River at mile 65.5. A variety of prehistoric and historic cultural habitation and use areas have been documented across the delta. In 2009 extensive excavations of C:13:0099 documented the presence of a small prehistoric room (possible kiva) with a central masonry feature (Feature 12 on GRCA site form). Once excavations were completed, the site was backfilled with sand and recontoured with the internal feature still intact. Since a buried structure containing a central feature was known to be present, the examination of this locale presented a good blind test of the effectiveness of terrestrial remote sensing techniques as a management tool along the Colorado River (Mink 2013). Both GPR and magnetic gradiometer surveys were conducted at C:13:099 with the following results.

The GPR data from C:13:0099 were collected within a 7 x 10 m grid that encompassed the reburied kiva. The results (Figure 4.5a) identified several interesting anomalies within 60 cm of the surface, with two of the uppermost slice maps (a3–a4) exhibiting a large, amorphous, low-reflection anomaly that is represented by the lighter tones around the darker anomaly. These anomalies represent the structural depression (light-colored anomaly) and central feature (dark-colored anomaly). It should be noted that since Feature 12 was backfilled with clean sand and other obstructions were removed during excavation, the radar waves behaved differently than if the site had been naturally filled in, creating a much clearer void than would be visible in an unmitigated site.

The magnetic gradiometer data (see Figure 4.5b) at C:13:0099 were collected within a 10 x 10 m grid that also encompassed Feature 12. The gradiometer survey data did not identify as many anomalies as the GPR survey. In particular, the outlines of the structure are not as apparent, though the central feature is represented by a positive monopolar anomaly that has a stubby *T* shape. Several other monopolar anomalies represented by positive magnetic reflections (dark spots) also may represent internal archaeological features.

Overall, the geophysical survey results at C:13:0099 were constructive. The central feature within the structure and the outline of the backfilled structural depression were clearly discernible in the GPR results. As this site was previously excavated, it provided a good test to determine how well the geophysical equipment performed in backcountry Colorado River corridor conditions. As the cause of these anomalies is known, these data would be a prime case study to include in a terrestrial remote sensing anomaly signature database.

## B:16:105

Site B:16:105 is the first archaeological site within Grand Canyon National Park to be investigated utilizing geophysical techniques. The site is a small-settlement locale that encompasses the surface remains of a small masonry-outlined pit structure and a stone-outlined cist. Based on the ceramic arti-

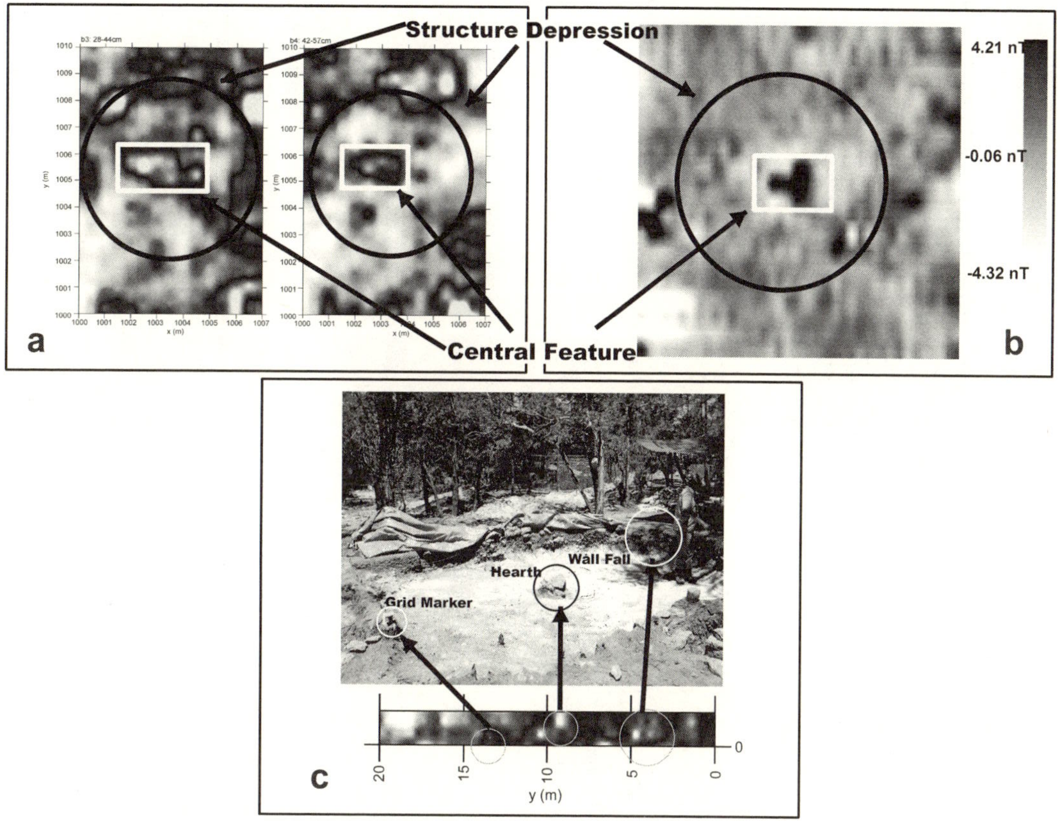


Figure 4.5. Results of (a) ground-penetrating radar survey at C:13:0099, (b) magnetic gradiometer survey at C:13:0099, and (c) ground-penetrating radar survey at B:16:105.

facts, the site dates to between the mid-eleventh century and the early twelfth century (Downum 2010). The site was excavated as a mitigation measure related to the construction of nearby Park Service housing. Prior to the excavation I was given a couple of hours to conduct a small GPR survey over the masonry-lined pit structure. The results of this survey (see Figure 4.5c) clearly show the outline of the pithouse along with several internal features including the hearth. Because this site was excavated, the results provide valuable data on what future geophysical results at similar sites might represent. Conducting geophysical surveys on sites slated for excavation for mitigation purposes provides a valuable set of baseline data that can inform future interpretations of sites and be used in the development of a terrestrial remote sensing anomaly database. Establishing an anomaly signature database is critical in a region where so few terrestrial remote sensing survey results have been published.

## Conclusion

The terrestrial remote sensing data collected at sites throughout the Grand Canyon have contributed to the understanding of Pueblo period settlement in the park. While only a limited number of sites have been examined, results from these investigations have confirmed some hypotheses while challenging others. The results from sites C:13:0778 and MU 3617 seem to confirm Schwartz's Biseasonal Settlement model and expand it to the South Rim populations, as suggested by the Cross Canyon model. The data from C:13:0778 also indicate that the shifting settlement patterns along the Colorado River below the canyon rim proposed by Anderson and Neff (2011) may not apply throughout the river corridor. Additional geophysical surveys at sites within the park will only continue to add to our understanding of Pueblo period settlement practice at Grand Canyon.

In addition to aiding a researcher's academic inquiry, terrestrial remote sensing can also contribute to federal land management research and responsibilities. NPS cultural resource managers are often forced to juggle the dual intentions of legislative mandates, such as Section 110 of the National Historic Preservation Act. Section 110 requires not only that resource managers preserve the cultural resources under their care but that they also nominate these resources for the National Register of Historic Places. Nominations require detailed data about a site to determine its importance to our knowledge of the past. Technologies such as terrestrial remote sensing provide cultural resource managers with a set of tools to meet the dual needs of preservation and data acquisition with little or no disturbance to archaeological and environmental resources. Moreover, the noninvasive nature of these techniques makes them a culturally sensitive tool to acquire archaeological data without disturbing burials or other features regarded as sacred by indigenous peoples. However, to ensure that these techniques continue to be useful, we must do a better job at developing tools such as anomaly signature databases to guide us in making the appropriate inferences about the cause of an anomaly, which in turn expands our ability to interpret a site in a larger context.

All the data presented in these case studies have not only advanced my research but have provided Grand Canyon National Park with valuable information on subsurface archaeological deposits to help better manage the resources under its purview. These cases illustrate the importance of utilizing geophysical surveys in twenty-first-century Southwestern archaeology. Haury's premise that data are the best resource in an academic confrontation still holds true, and terrestrial remote sensing data are another armament in the arsenal. It is time in Southwestern archaeology for the tip of the shovel to be not replaced, but enhanced, by the tip of a GPR amplitude wave.

## Acknowledgments

I thank the cultural resource program staff at Grand Canyon National Park, in particular Ellen Brennan, Jen Dierker, Amy Horn, and Jan Balsom, for their intellectual and administrative support for my research. I would also like to thank Alan Sullivan of the University of Cincinnati for his constant support and for allowing me to collect data at MU 3617. Portions of the research were carried out under a cooperative agreement between the University of Kentucky and Grand Canyon National Park through a task agreement managed by the Southern Appalachian Cooperative Ecosystem Studies Unit (TA #H000095041). Geophysical equipment was purchased by the University of Kentucky with a National Science Foundation Major Research Instrumentation Grant (#0619464).

# 2

# Technological Transformation and Economic Change

# 5
# A Remote Sensing Approach to Studying Land Use Change in Chaco Canyon, New Mexico

*Jennie O. Sturm*

The ways in which humans use and modify their landscapes often reflect larger social, economic, and political changes within a societal system. In North America in general, and the American Southwest in particular, land modification related to prehistoric land use has reemerged as a topic of interest because of its hypothesized long-term impact on ecosystem structure and implications for social complexity (Dean 1992; Van West 2011). Yet locating and characterizing the spatiotemporal patterns associated with prehistoric land use are often major methodological challenges. Unless the land use activity included substantial architecture, many features related to land use in the Southwest can be easily buried over time, or can be so subtle that they are hard to recognize as such. Furthermore, detecting *change* in land use, or how a possible landscape was modified over time, can be even more challenging.

Remote sensing, especially when combined with other geospatial methods, offers a means of detecting land use patterns that may otherwise be excluded from the cultural interpretations of an area. While remote sensing has been used in American archaeology for many decades, the ways in which these methods are applied to problems, and the interpretations that are now possible from them, continue to expand and evolve.

This chapter presents a case study using remote sensing methods to distinguish complicated sets of localized land use patterns in Chaco Canyon, New Mexico. Prehistoric land use in Chaco is a complex topic with differing interpretations, and this is exemplified by a feature often referred to as the "Chetro Ketl field" (Figure 5.1). With a time series of historical aerial photographs and ground-penetrating radar (GPR), complex configurations of features buried below the visible part of the Chetro Ketl field were identified. While the function of these buried features is unclear, the results from this study do suggest that this overall area was rebuilt at some point, possibly reflecting an economic or social shift in Chacoan society. Further-

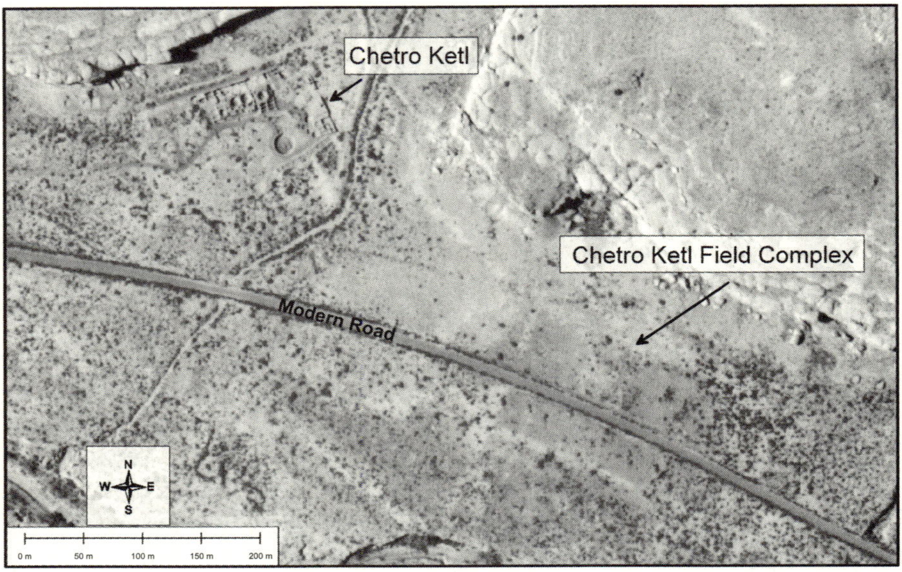


Figure 5.1. Chetro Ketl field, located southeast of the Chetro Ketl great house in Chaco Canyon, New Mexico.

more, it is likely that other extensive buried complexes of features that are not evident on the surface exist in additional parts of the canyon and beyond. The remote sensing approach developed for this specific study can provide a nondestructive strategy for locating them, offering new insights about the "invisible" Southwestern landscape.

## Background

Chaco Canyon, located in northwestern New Mexico in the American Southwest (see Figures I.1 and 5.1), hosted one of the largest and most socially complex societies of the Ancestral Puebloan world. Flourishing from about A.D. 850 to 1150, Chaco is best known for its monumental, multistory great houses, numerous great kivas, highly formalized masonry building style, elaborate ceramic designs, and imported exotic materials (Judge 1991:24). Today, Chaco Canyon is administered by the National Park Service (NPS).

Chaco's great houses are often cited as some of the clearest examples of emergent social complexity in pre-Hispanic North America and therefore have been researched extensively (e.g., see Lekson 2007; Neitzel 2003). Far less attention has historically been given to the land in between these great houses, leading to ambiguity about what types of land modification occurred and how that related to land use activities. The research that has been done has led to conflicting interpretations regarding land use, even when the same evidence is used.

For example, some researchers have argued that Chaco's landscape served a primarily economic function by being modified to control the flow of water onto agricultural fields (Sebastian 1992; Vivian 1970, 1974, 1990; Vivian et al. 2006). The example most often cited as an agricultural field is often referred to as the "Chetro Ketl field." This area is a large (roughly 2 ha), rectangular feature a few hundred meters southeast of the Chetro Ketl great house (see Figure 5.1). In addition to the overall differences in soil coloration and vegetation, a faint interior grid pattern is visible in most aerial photos. Gridded fields, in which a larger area is subdivided into smaller plots, are well-documented features in the agricultural economies of many protohistorical and historical Pueblo societies (Anschuetz 2006; Dominguez 2002; Maxwell and Anschuetz 1992; Wills et al. 1990). It has been hypothesized that if one of these gridded fields exists in Chaco, then others likely do as well and are simply buried under alluvial deposition (Vivian et al. 2006).

Not all researchers, however, agree with the "agricultural landscape" interpretation. A common alternative explanation of land modification at Chaco relates to what has been called a sacred or ritual landscape (Van Dyke 2007). This perspective explains many elements of the built environment as establishing sacred connections to a worldview, promoting ceremonial activities, or reinforcing social ideology (Sofaer 2007; Stein et al. 2007). For example, under this explanation, the "Chetro Ketl field" is interpreted as a series of ball courts or areas for ceremonial performances (Stein et al. 2007:211).

Both of these examples of interpreting prehistoric land use at Chaco include evidence from the same feature (the Chetro Ketl field), making it one of the most ambiguous instances of land modification in the canyon. Each interpretation of this feature carries vastly different implications for Chaco's societal structure, including agricultural infrastructure and innovation, economic intensification, resource management, and ritual investment. Therefore, refining how the Chetro Ketl field is interpreted directly affects our understanding of the potential causal relationships between Chacoan society and the resulting land modification.

One major reason the Chetro Ketl field has been so difficult to interpret is that no obvious comparable features in Chaco have been located (Love 1980; Morain et al. 1981; Stein et al. 2007; Stein and Fowler 1996). Concerted efforts were made in the 1970s and 1980s, when a number of remote sensing technologies were applied in Chaco. Many of these experiments, most notably those using aerial photography and photogrammetry, were pivotal in Chacoan archaeology, helping to locate and define features that were imperceptible from the ground surface. In addition to aerial techniques, a number of geophysical methods, including GPR and magnetometry, were used in an attempt to map buried features in Chaco. These efforts, which included experiments at Pueblo Bonito and the Chetro Ketl field, were some of the first pioneering applications of geophysics in American archaeology (e.g., see Lyons and Avery 1977; Vickers et al. 1976). Despite some promising results, most of these geophysical efforts were inconclusive. However, it is noteworthy that rather than abandon the promise of such methods, researchers framed the problem in terms of future technological development, hypothesizing that "refinements in . . . instrumentation and data interpretation programs may allow detection of subtle soil density changes typical" of these features (Loose and Lyons 1976:145).

Forty years later, this statement seems almost prophetic. Geophysical methods that were once experimental have become standard practice in many types of archaeological fieldwork, including in the American Southwest. Advancements in technology and computing power mean large areas can be covered quickly, and data can be processed into images that are readily interpretable (Conyers 2013). Refocusing remote sensing efforts in Chaco not only carries on the important legacy of such work started in the 1970s but also offers a way to study land use in Chaco while adhering to the NPS mission of preservation.

As shown in the following sections, remote sensing and geospatial methods offer a logistical solution to studying land use patterns in Chaco noninvasively, while also generating data about the Chetro Ketl field that are unobservable from any other source. The maps produced from this study are both independent of previous archaeological studies and able to be used to guide and complement any future excavations here.

## Methods and Approach

In this project, GPR was used as the primary method for mapping buried features within the Chetro Ketl field. Using GPR in conjunction with other geospatial methods helped contextualize the GPR results with the surrounding landscape. These methods included a time series of historical aerial photos that were used to document and define the visible surface patterns in this area, a real-time kinematic global navigation satellite system (RTK GNSS) to accurately relate the GPR survey areas to the aerial photos, and integration into a GIS environment to allow a platform for analysis and interpretation.

Chaco Canyon has been photographed for decades, and a multitude of

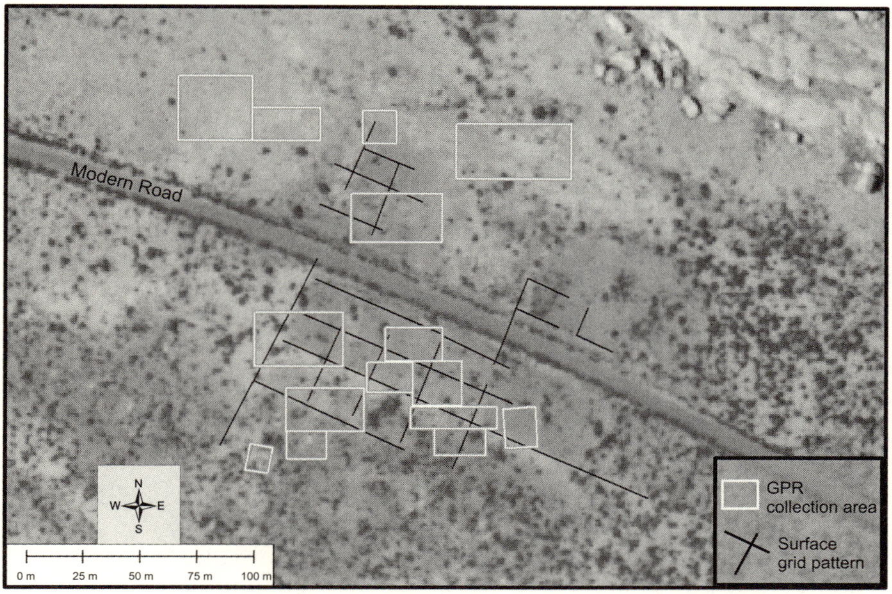


Figure 5.2. Close-up of the Chetro Ketl field. The linear features visible in a time series of historical aerial photos have been digitized with black lines. The GPR survey grids (white rectangles) are also noted.

aerial photos dating back to the 1920s exist for the area. While these photos have varying degrees of resolution, the time depth they offer is an excellent way to account for some of the landscape changes of the past century (Jensen 2005). For the Chetro Ketl field, historical aerial photos are the only way to observe features that existed prior to NPS building a modern road directly through the field (see Figure 5.1).

With six sets of photographs ranging in date from 1935 to 2011, interior plot borders within the Chetro Ketl field were digitized and recorded. Only those plot borders that were clearly visible were recorded, without inferring borders that might have created a full grid. All photos used for this analysis are housed at the Earth Data Analysis Center (EDAC) and online GIS repository at the University of New Mexico, and each was georeferenced and coregistered using the most recent aerial photos (1 m resolution) as a reference frame. With this strategy, numerous plot borders were recorded that are not apparent on contemporary imagery, including some that existed prior to the modern road being built. The final product of this effort is a composite image showing all borders visible on the surface over the course of nearly a century (Figure 5.2).

Once the aerial photo analysis was complete, sample areas within the field were selected for GPR survey. While the overall field area is easily visible in aerial photos, it is almost imperceptible on the ground, a problem noted by geophysical archaeologists in the 1970s as well (Loose and Lyons 1976:134–137). To overcome this visibility problem, all GPR grids were selected for placement based on the aerial photos in a GIS environment. These survey areas were then made into rectilinear grids and sized to maximize the amount of ground covered while also limiting the amount of vegetation included. These survey grids were then input as a shapefile into the RTK GNSS to stake out on the ground, thereby allowing specific areas of interest to be targeted and eliminating any guesswork from the survey location. Perhaps most importantly, this strategy easily allowed the GPR grids to be oriented so they were offset from the linear plots visible in aerial photographs (see Figure 5.2). This helped avoid collecting transects parallel to the linear plots, which would run the risk of being removed as horizontal banding during postprocessing (Conyers 2013).

To minimize the impact on vegetation within the park and maximize available field time, survey areas were chosen in locations that were generally free of large bushes and other major surface obstacles. The majority of GPR grids were placed over the visible grid borders with the goal of better defining the field and testing for buried features. Two GPR grids were placed outside these visible grid borders to test for features outside the grid area (see Figure 5.2).

The GPR method was selected for this study because of both previous contemporary successes using it within the canyon (e.g., see Sturm and Crown 2015) and the need for high-resolution maps that could be analyzed at various depths in the ground. Over the course of three field seasons, GPR data were collected with a GSSI SIR 3000 controller using 400 MHz antennas. All data within the field were collected in rectilinear grids using 50 cm transect spacing. The data were processed to generate both reflection profiles and amplitude slice maps using the programs GPRViewer+, GPR Process, and GPR-SLICE. Various standard processing techniques were utilized in an iterative process to generate interpretable maps, including background filtering, low-pass filtering (3 x 3), and a high data sampling rate to generate relatively thin slice maps (2–3 nanoseconds) (Conyers 2013; Goodman and Piro 2013). To convert time measurements into actual depth, a velocity analysis was performed (Conyers 2013; Conyers and Lucius 1996) and resulted in an average relative dielectric permittivity (RDP) of 5.2.

Fifteen grids have been collected thus far. All the amplitude slice maps that were generated for each survey area were converted to raster images and georectified to the aerial photos in a GIS environment. Features visible in the slice maps were then digitized to compare to the plot borders identified in the historical aerial photo analysis. This allowed a platform for visualizing how subsurface features are related to surface features.

## Results

A number of shallowly buried, long, linear features were identified in many of the GPR slice maps (Figure 5.3). Some of these linear features have a low reflectivity and appear as only subtle reflections in the amplitude slice maps. Still, the linear nature of these reflections is unmistakable (see Figure 5.3). To determine possible association, the linear features identified in the slice maps were analyzed according to depth after being converted from time measurements. While all the identified linear features appear shallow (occurring within the first 50 cm of subsurface), the depth analysis showed that some linear features consistently occurred below others. In total, three distinct categories of linear features were identified and interpreted from the GPR maps: (1) the grid borders visible in aerial photographs, occurring at the surface and extending to about 8 cm in depth (Figure 5.4b); (2) linear features not visible in aerial photographs, but oriented in the same direction as the surface grid pattern (these were often found within the grid borders visible in aerial photos and were mapped primarily at a depth of 10–20 cm; see Figure 5.4c); and (3) linear features neither visible in aerial photographs nor oriented in the same direction as the surface grid borders. These features were mapped primarily at depths ranging from about 15 to 50 cm (see Figure 5.4d). In addition to the linear features, a number of shallow, basin-like circular features were mapped that also appear primarily at depths from about 10 to 50 cm (see Figure 5.4d). All these circular features appear in the southern part of the field, and none appear to intersect the deeper linear features. These basins range in size from about 2 to 5 m in diameter, and the majority appear to extend to no more than 90 cm deep.

## Discussion

The results from this study show that there are configurations of buried features in the Chetro Ketl field that are distinct from the grid patterns apparent on the present ground surface (Figure 5.5). The linear features visible on the surface are oriented approximately 45 degrees from the buried linear features and are spaced about 20 m apart at their widest, which is twice the distance of the buried linear features. On the south end of the field,

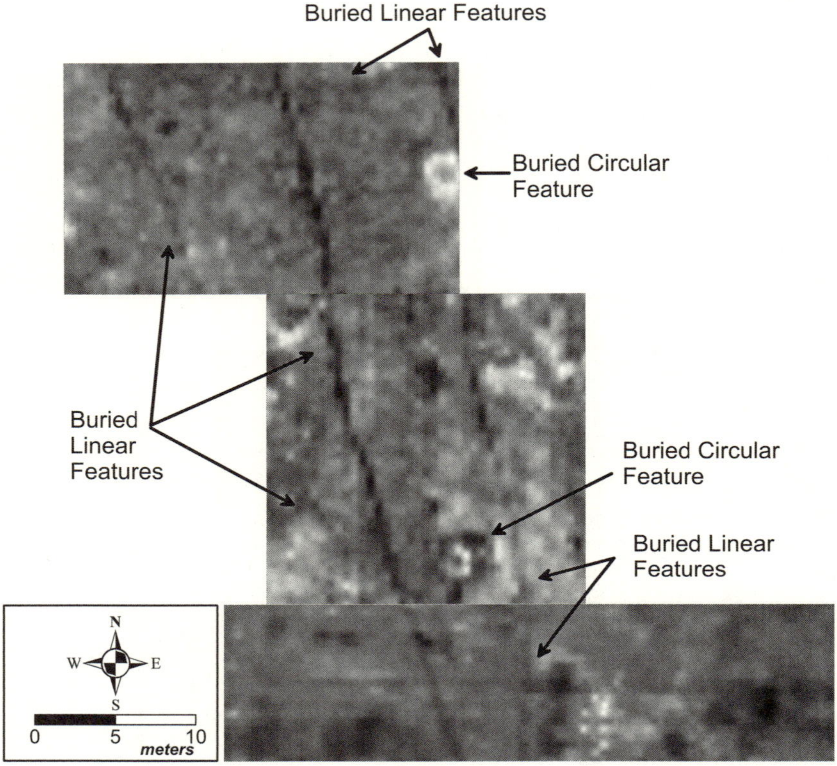


Figure 5.3. Three amplitude slice maps from three distinct grids in the southern part of the field. The reflections from several buried linear features and two circular features are apparent in the grids, and at least one linear feature extends through all three grids. These linear features are buried below the grid pattern evident in aerial photos, suggesting that they are associated with an earlier feature. These slice maps represent a depth of approximately 12–20 cm.

the buried linear features become narrower and cross each other in at least one place (see Figure 5.5). The southern part of the Chetro Ketl field is also where the buried circular features are located. None of these circular features overlap with or intersect the deeper linear features, suggesting that they were either already there or were subsequently built within or in conjunction with the original linear features. While the specific function of these circular features is currently unknown, their presence and patterning with the deeper linear features suggest that they may be part of the same configuration.

Excavation records from testing within the Chetro Ketl field during the

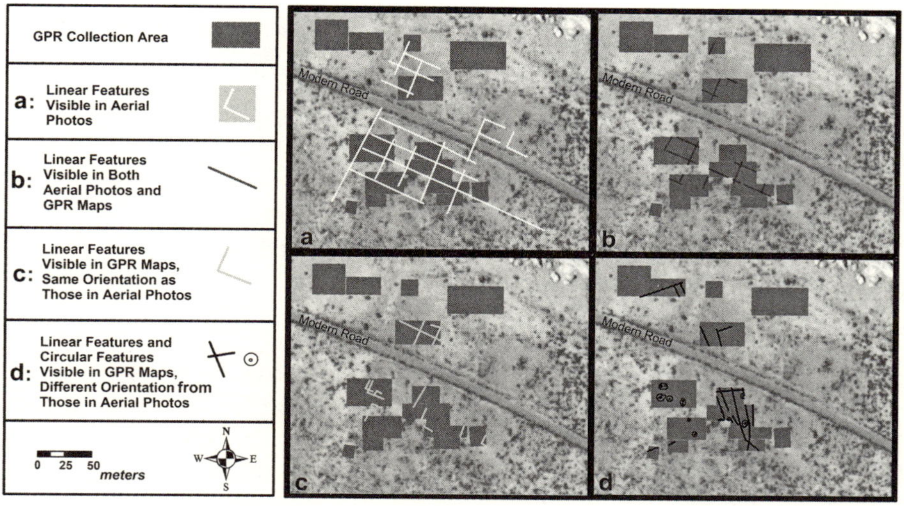


Figure 5.4. Chetro Ketl field showing (a) linear features identified from a time series of historical aerial photos; (b) linear features that were mapped with GPR and are visible in the aerial photos; (c) linear features that were mapped with GPR but are not visible in the aerial photos yet are oriented in the same direction; and (d) linear features mapped with GPR that are not visible in the aerial photos or oriented in the same direction. Also shown are the GPR survey grids and a number of small circular features that were mapped with GPR.

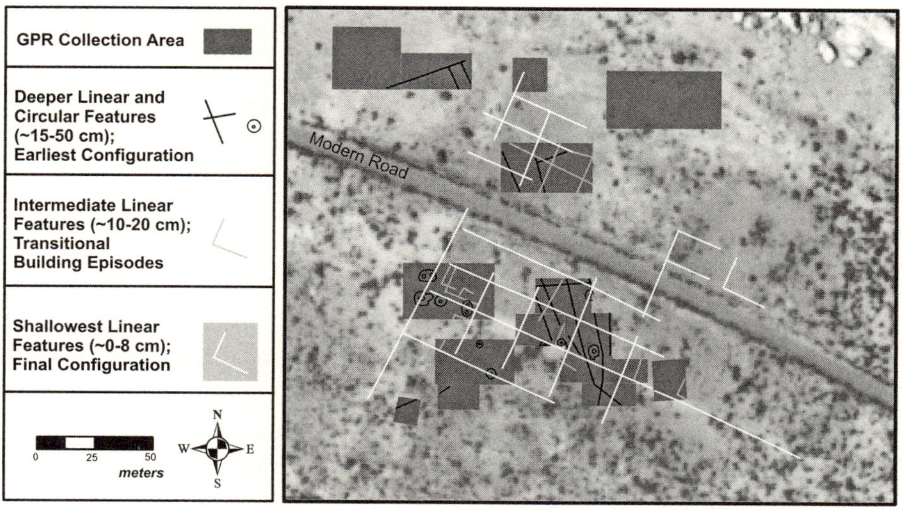


Figure 5.5. Chetro Ketl field showing the hypothesized relative chronological history of this area. The deeper linear features likely represent an earlier configuration. At some point, new linear features were built, rotated, and superimposed directly on top of the earlier configuration. All that is visible in aerial photos today is the final grid pattern.

1970s offer some clues to the nature of these buried linear features. These records show the presence of earthen berms interspersed with sand and clay layers throughout the Chetro Ketl field (Loose and Lyons 1976). In the GPR reflection profiles, these linear features are often difficult to distinguish from their surrounding sediments, but most appear as subtly reflective vertical "breaks" in the subsurface. This suggests that if these are earthen berms, then they may have been constructed more like walls with vertical edges than mounded berms with sloped sides.

Taken together, the information derived from the GPR analysis, historical aerial photo analysis, and GIS integration appears to represent a temporal series of building episodes. The earliest (deepest) evident feature consists of long, linear features oriented northwest to southeast and spaced approximately 10 m apart, with perpendicular linear features spaced approximately every 55 m. These linear features become narrower and less uniform in the southern part of the field complex, crossing in at least one place. Also possibly associated with this earlier configuration are the multiple circular basin features.

At some point, new linear features were built in this area and superimposed directly on top of the earlier system. The entire layout appears to have been rotated about 45 degrees from its original configuration and structured to form grids that measure approximately 23 m east-west and 14 m north-south. These grids are remarkably uniform in both shape and size. It is also possible that intermediate features were built during the reconstruction of this area, as evidenced by other partial linear features buried within, and oriented to, the surface grids. All that remains visible in aerial photos, however, is the final gridded pattern (see Figure 5.5).

## Conclusion

The Chetro Ketl field is one of the most ambiguous examples of land modification in Chaco Canyon because of its seemingly unique presence. The grid pattern visible on the surface has been interpreted to represent very different land use functions, from a gridded agricultural field to ceremonial performance areas. None of these interpretations, however, has been conclusively linked to its functional correlate. The results from this study have shown that what is visible on the surface is actually only one component of the Chetro Ketl field, and this may be one of the reasons the Chetro Ketl field has been so difficult to interpret. Therefore, while this study does not necessarily establish its function, it does suggest that there is a significant component of this feature currently being excluded from its interpretation. It is possible that no part of this feature would be visible today if it had not undergone extensive remodeling at some point in its history. This suggests the tantalizing possibility that other land use patterns exist throughout the canyon but remain invisible from the surface because of sediment deposition. More work needs to be done to expand our understanding of this feature and determine whether it fits within a broader system of land use in Chaco or represents a localized palimpsest of prehistoric activity.

Determining that an extensive pattern of buried features exists below the gridded pattern seen in aerial photos could have been accomplished only with GPR survey. The field area and its gridded plots are imperceptible on the ground, and none of the buried features located are identifiable in aerial photos. The ability to analyze depth from the GPR data meant it was possible to recognize the buried features as likely associated with each other. Still, using GPR in combination with other geospatial methods, including the time series of historical aerial photos, a subcentimeter GPS (RTK GNSS), and GIS, truly allowed the most robust analyses. The accurate integration of the information derived from multiple mapping platforms made it possible to recognize this episode of land modification and hypothesize a relative chronological sequence of building activity for this area.

In addition, these buried features are so expansive that they would not be detectable with excavation without stripping nearly two hectares of ground surface. Excavations in the 1970s successfully revealed some of the complex stratigraphy in this area, but the excavators had no way of knowing that some of what they observed related to a different configuration of features, let alone the size, orientation, or expanse of that original complex. This study therefore stands in stark contrast to misconceptions that remote sensing is an initial estimate of features in an area that then needs to be refined through excavation. Rather, this study shows that remote sensing can be used as a stand-alone approach to generate new data that can be used directly to test an archaeological question in ways that are not possible using other sources. Should excavation occur here in the future, these maps can now effectively guide the placement of those excavations in ways that are more thoughtful and efficient.

The use of remote sensing in American archaeology is not new, but the ways in which archaeologists apply these methods and integrate them with other technologies will continue to evolve. New questions and problems will require new approaches and solutions. In a place like a national park, where preservation and protection of cultural resources are the park's primary missions, the need for noninvasive or minimally invasive archaeology will continue to grow in importance (see also chapter 4 in this volume for another example of this concern in national parks). The use of remote sensing not only helps overcome logistical problems with studying certain archaeological sites but also expands the types of questions that can be asked. Given that the alternative might be to avoid working in such areas altogether, the use of remote sensing technology can only be considered beneficial for archaeology.

## Acknowledgments

This work was conducted as part of a mutually funded cooperative agreement between Chaco Culture National Historical Park (CCNHP) and the University of New Mexico (UNM). I would like to gratefully acknowledge the personnel at CCNHP, especially Dabney Ford and Roger Moore. Thank you to W. H. Wills at UNM for indispensable guidance on this work. Pablo Flores and a number of student volunteers provided valuable field assistance during data collection. Thank you to Duncan McKinnon and Bryan Haley for organizing this volume and for providing helpful suggestions. Three anonymous peer reviewers also provided valuable observations. Finally, thank you to all the pioneers of geophysical archaeology, especially my great friend and mentor Larry Conyers. We are but standing on the shoulders of giants.

# 6
# Examining Agricultural Surplus at Huff Village, North Dakota

## Combining Archaeological and Remote Sensing Data

*Adam S. Wiewel*

Archaeologists around the world have been more than awestruck by the impressive results of remote sensing investigations in recent decades. Extensive geophysical surveys in locations such as the Roman city of Wroxeter (Gaffney et al. 2000), the Iron Age site of Kerkenes in Turkey (Atalan et al. 2008), the prehistoric earthworks of Ohio (Burks and Cook 2011), and even Plains Village period settlements in North Dakota (Kvamme and Ahler 2007) have yielded invaluable insights that would otherwise not have been attainable without destructive archaeological excavations. Such undertakings are increasingly important to the discipline. Large-scale excavations in North America and many other parts of the world have become uncommon for both economic and ethical reasons. Remote sensing and geophysical investigations may therefore serve as a bridge between traditional pedestrian surveys of landscapes and site-level data generated by excavations.

In recent years the practice of archaeological geophysics has been aided by improvements in instrument reliability, data acquisition speed, software processing capability, and visual display techniques. Despite these advances and the widespread availability of geophysical equipment, archaeological feature discovery remains their primary use (V. Thompson et al. 2011). While this use is crucial to the discipline, especially in cultural resource management contexts (see chapters 2 and 4 of this volume), archaeological geophysicists often stop short of the more difficult task of addressing anthropological questions. Yet in many circumstances geophysical surveys may yield primary information suitable to address such questions (Kvamme 2003a). To this end, I directly pose questions concerning the number of storage pits, total storage capacity, population size, agricultural production, and per capita storage capacity at Huff Indian Village State Historic Site (32MO11) by combining magnetic gradiometry and high-resolution elevation data with coring and excavation information.

## Huff Village

Huff Village is a Plains Village tradition fortified settlement located on the west bank of the Missouri River in present-day North Dakota (see Figure I.1). Radiocarbon dates indicate the village was occupied by ancestral Mandans for a brief period during the mid-fifteenth century (ca. A.D. 1440–1460) (Ahler 2000). The site has been the subject of archaeological inquiry for over a century (J. Howard 1962; Will and Hecker 1944:19–23, 94–96; Wood 1967:24–28). The earliest investigations were aimed primarily at mapping the extent of features visible on the ground surface (Wood 1967:26). The most comprehensive study at Huff was undertaken by W. Raymond Wood in 1960. His salvage work focused on a strip approximately 30 m wide along the cut bank of the Missouri River (Lake Oahe) in conjunction with leveling and bank stabilization efforts. This investigation included (1) excavations of portions of the fortification ditch and palisade line; (2) excavations of eight houses; (3) exposure, complete excavation, remapping, and description of three previously excavated houses; and (4) accurate mapping of the site with the aid of a vertical aerial photograph and repetitive field checks (Wood 1967:Map 4).

The village, surrounded by a fortification system on three sides and bordered by the Missouri River on the fourth, is roughly rectangular, with an area of approximately 4.8 ha (48,301 $m^2$; area calculated within the outside edge of the fortification ditch based on remote sensing data and Wood's map). The fortification system, approximately 650 m in total length, includes a shallow ditch, a palisade wall, and 10 regularly spaced bastions. Wood (1967:Map 4) mapped 103 surface depressions indicating the locations of houses, and excavations revealed that most are rectangular, although at least one exhibits four-post rounded square architecture. The houses were roughly constructed in rows that parallel the river with their entryways facing the southwest. An open area or plaza toward the village center is indicated by its notable lack of house depressions.

Importantly, Wood's (1967) work at Huff Village played a crucial role in the definition of the Huff focus or Terminal Middle Missouri variant, an archaeological taxon that represents the final stage of the Middle Missouri tradition, of which Huff is the type site (Ahler 1993; Lehmer 1971). Among many differences with preceding settlements, archaeologists have noted changes in settlement patterning during the mid-fifteenth century, especially an increase in settlement density and a decrease in settlement number (Lehmer 1971:120–128; Mitchell 2013:67–71, 170–173). Additionally, limited excavation data indicate changes in agricultural productivity beginning in the fifteenth century and continuing into the eighteenth century (Lehmer 1971:140; Mitchell 2013:178–179). Specifically, it is thought that the total number and size of subterranean food storage pits within Middle Missouri Plains Village sites increased through time (Mitchell 2013:178–179). Pits were used to store agricultural produce such as beans, sunflowers, squash, and especially corn for later consumption and trade (G. L. Wilson 1917:87). Quantitative data regarding their quantities and volumes within sites have clear and important implications for village occupants' capacity to produce an agricultural surplus and for their participation in intertribal trade. However, few systematic attempts have been made to determine the number of storage pits, total storage capacity, and per capita storage capacity within any Plains Village settlement. Short of completely excavating entire sites, one is left to use alternative data sets to yield these estimates. I use excavation and coring data along with the results of extensive magnetic gradiometry and topographic mappings at Huff Village for this purpose.

## Storage Pits at Huff Village

Within the 10 houses described in detail by Wood (1967:31–52) (approximately 10 percent of the total number mapped at Huff Village in 1960), 48 straight-sided or cylindrical pits and undercut or bell-shaped pits (excluding post molds, hearths, basin-shaped features, or small "pocket caches" less than 0.1 $m^3$ in volume) were documented. The excavated houses contained as few as one and as many as nine cylindrical and bell-shaped pits, found lining the structures' walls, near their corners, and within or close to entryways. I estimated the volume of bell-shaped pits with an equation for the volume of a conical frustum [$V = (1/3) * \pi * h * (r_1^2 + r_2^2 + (r_1 * r_2))$, where $r_1$ is the lower radius, $r_2$ the upper radius, and h the height], while I calculated straight-sided pit volumes with an equation for the volume of a circular cylinder ($V = \pi r^2 h$). Estimated pit volumes vary from a low of 0.1 $m^3$ to as much as 4.8 $m^3$, with a mean pit volume of 1.03 $m^3$ ($s$ = 0.83 $m^3$) and a total storage capacity of 49.6 $m^3$. With a total of 103 houses, this would suggest approximately 494 pits across the site and a total storage capacity of nearly 509 $m^3$. An apparent problem with this estimate is that it is based entirely or almost entirely on storage pits excavated within a 10 percent nonrandom sample of dwellings. A pilot magnetic gradiometry survey in 1999 (Kvamme 2003b; Kvamme and Ahler 2000) demonstrates that this sample is biased.

## Previous Geophysical Surveys at Huff Village

Kvamme's (2003b; Kvamme and Ahler 2000) initial geophysical investigations at the site were performed with the goal of sampling a variety of the site's features, including its fortification system, lodges, and plaza. The largest survey was by magnetic gradiometry and covered approximately 0.71 ha. Although the survey area covered only a portion of the entire village (approximately 15 percent), it revealed several significant insights. Most importantly, the gradiometry survey indicated that many more likely storage pits were located *outside* rather than *within* houses, particularly in the spaces surrounding each house (Kvamme 2003b:136–139; Kvamme and Ahler 2000). To aid interpretation of the considerable number of magnetic anomalies, systematic coring in a 900 $m^2$ block encompassing multiple dwellings, part of the plaza, and an area devoid of houses was performed immediately after the survey (Kvamme and Ahler 2000:16–33). Within this area, mapped magnetic anomalies were found to be more common in outside-house locations (density of 9.4 versus 6.4 per 100 $m^2$). This is an important consideration given that excavations had previously focused on within-house contexts. Additionally, magnetic anomalies within houses appeared to be significantly smaller (one-third to one-quarter the diameter of those that were outside dwellings).

Thirty-five discrete magnetic features were cored and identified as storage pits ($n$ = 26, or 74 percent), hearths ($n$ = 5, or 14 percent), middens ($n$ = 2, or 6 percent), or indeterminate ($n$ = 2, or 6 percent). Using these figures, Kvamme and Ahler (2000:33–34) estimated a total of 2,046 pit features, 386 hearths, and 165 middens across the entire village. Given an approximate volume of nearly 1.2 $m^3$ (based on a standard pit depth of 1.5 m and a diameter of 1 m), the total storage capacity of the village would have approached 2,500 $m^3$.

Interestingly, Kvamme and Ahler utilized a larger average storage pit volume (1.2 $m^3$) than that derived from Wood's excavated pits (1.03 $m^3$), a difference that may have significant implications given the site's many storage pits. This difference may relate to the apparent dissimilarity in storage pit sizes within and outside houses noted by Kvamme and Ahler, although their observation was based on the diameter of magnetic anomalies rather than the actual size of excavated features. Wood's excavation data from 48 storage pits are enlightening. While the mean orifice diameter of 1 m agrees with Kvamme and Ahler's diameter, the data reveal a significantly different mean base diameter of 1.2 m ($t$ = 4.579; $df$ = 47; $p$ < 0.001). Moreover, Wood's mean pit depth is only 0.93 m, significantly different from Kvamme and Ahler's postulated 1.5 m ($t$ = -12; $df$ = 47; $p$ < 0.001), but Kvamme and Ahler's own coring data show a mean pit depth outside houses of only 0.95 m, not significantly different ($t$ = 0.195; $df$ = 69; $p$ = 0.85) from Wood's data (and in disagreement with their own postulated value). In light of these results, an average storage pit depth of just less than 1 m and a volume of approximately 1 $m^3$ is likely a better estimate for storage pits at the site. Moreover, this difference would indicate a total storage capacity closer to 2,100 $m^3$, considerably less than Kvamme and Ahler's initial estimate of nearly 2,500 $m^3$.

## 2009 Geophysical Surveys at Huff Village

Because of the success of the initial limited survey of the site, expanded remote sensing investigations were undertaken at Huff Village in 2009 (Kvamme et al. 2009). The work was performed by Kvamme along with two graduate students, including me. The goal of this project was to completely survey the settlement, identify subsurface features with magnetic gradiometry, and document the site's surface topography with a robotic total station. This work provides an opportunity to estimate storage pit numbers and per capita storage capacity at the site.

## Methodology and Instrumentation

The magnetic gradiometry survey at Huff Village was performed with a Bartington Grad601, a dual-sensor fluxgate instrument. Archaeologists worldwide have found this method to be particularly fruitful because several natural and cultural processes alter soil magnetism (Aspinall et al. 2008: 24–25; Kvamme 2006b:205, 214–221). In the Northern Plains, this instrument permits hearths, storage pits, and other subsurface features to be detected because they vary in magnetism (Bales and Kvamme 2005; Kvamme 2007a). More than 5 ha were surveyed at Huff Village, completely covering the site and extending beyond its fortification system to the state park's fence (Figure 6.1).

Additionally, a 5.1 ha raster digital elevation model (DEM) of the entire site was produced at a 0.5 m spatial resolution with a Trimble 5600. This robotic total station has automatic aiming and tracking capabilities that enable position and elevation data to be continuously recorded (Kvamme, Ernenwein, and Markussen 2006). A local relief model was generated from the DEM by subtracting the linear trend of the ground surface (Figure 6.2a) (Kvamme et al. 2009). Finally, the topographic surface was better illustrated using sky view factor analysis, a visualization technique designed to illuminate topographic depressions by modeling "sky" occlusion (see Figure 6.2b) (Zakšek et al. 2011). These data sets were integrated in a GIS, and Wood's (1967:Map 4) map of Huff Village was registered and rectified, which aided comparison, interpretation, and digitization of past excavations and features of archaeological significance.

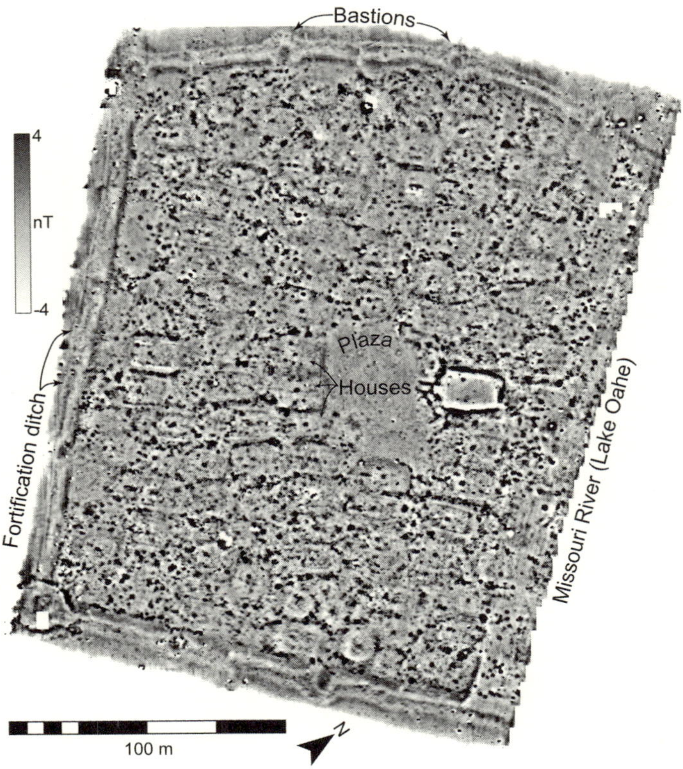


Figure 6.1. Magnetic gradiometry results at Huff Village.

## Interpretation of Features at Huff Village

Many of the features discussed previously are visible in both the magnetic gradiometry and elevation data sets (see Figures 6.1 and 6.2) (Kvamme et al. 2009). It is noteworthy that many features of likely archaeological significance are magnetically visible in the area of Wood's investigations adjacent to the Missouri River that was stripped by heavy machinery, suggesting they were not completely removed during the salvage operations. Likewise, house and bastion excavations are visible because the topsoil was removed and placed in piles next to the units. Additionally, the plaza is visible topographically as a relatively broad, level space near the village center that also exhibits few magnetic anomalies. The fortification ditch and bastions are quite apparent in both topographic and magnetic mappings. In particular, the distinctive magnetic signature of the fortification system is a product of its construction, where magnetically enriched topsoil was excavated from

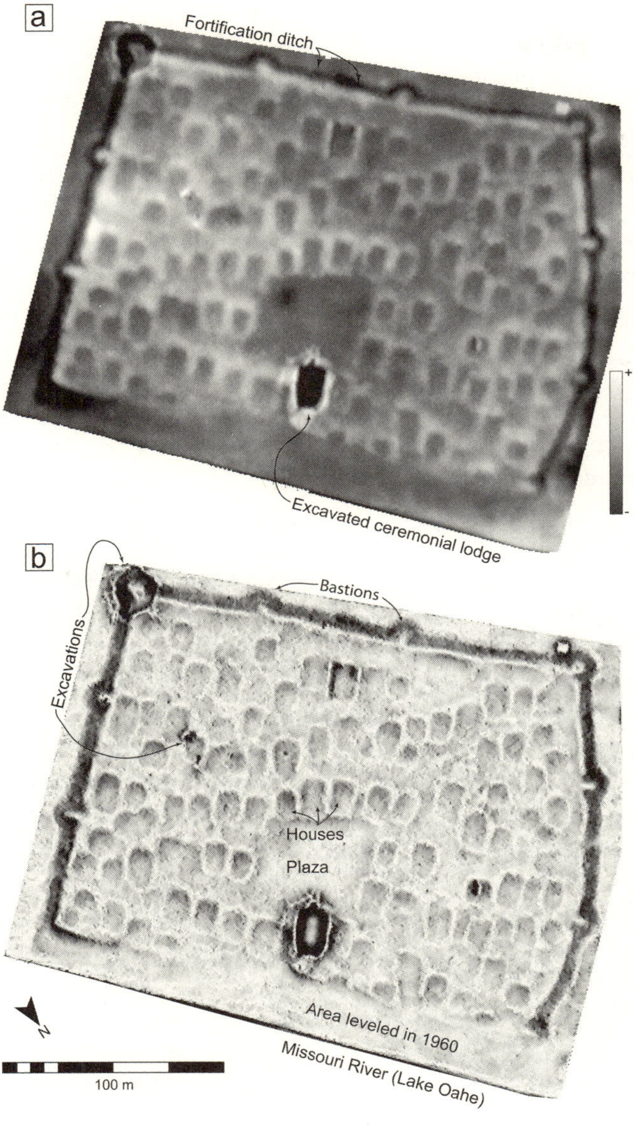


Figure 6.2. Topographic surface of Huff Village (a) following removal of the trend surface and (b) with estimation of the sky view factor.

the ditch and mounded inside it to form an elevated berm on which the palisade line was placed.

## Houses

The general form of each house is expressed both topographically and magnetically, and the data provide significant insights regarding the construction of houses at Huff Village (Kvamme et al. 2009). That is, the long, rectangular forms of most houses are visible in the elevation data because of their relatively level interior floor spaces and because the mounded soil that once made their walls now surrounds their perimeters (Figure 6.3a). The magnetic gradiometry data provide even more detail regarding houses, closely paralleling Wood's (1967) plan maps of excavated lodges (see Figure 6.3b). The magnetic data clearly indicate their forms as well as content (e.g., entryways, hearths, and interior storage pits). Also, the perimeters of many lodges are clearly visible as a narrow linear feature, likely the result of construction methods (e.g., the cutting and removal of sod and preparation of the level floor on the underlying soil).

## Discrete Magnetic Anomalies

Discrete magnetic anomalies approximately 1–3 m in diameter, roughly circular in shape, and with magnetic values ranging from about 2 to 20 nT (nanoteslas) are nearly ubiquitous across the settlement, with the exception of the central plaza. These anomalies are largely the result of modern metal debris, hearths, and storage pits. The first is relatively easy to identify because of its distinctive magnetic signature. That is, high-value magnetic anomalies indicative of recent iron or steel, including debris from past excavations, exhibit dipolar forms (paired positive and negative measurements). However, hearths and storage pits are more difficult to distinguish because they are roughly similar in size and magnitude. Anomalies indicative of the two feature types are generated by different processes (i.e., hearths form thermoremanent anomalies while pit features filled with magnetically enriched midden material result in induced anomalies). Yet they are not easily differentiated because magnetic gradiometers measure the sum of all forms of magnetism and are not capable of distinguishing them.

Previous studies of storage pits and hearths in Northern Plains villages have shown that hearths exhibit statistically greater magnetism than storage pits, although their distributions overlap considerably (Bales and Kvamme 2005). Still, reasonable interpretations may be made by drawing on a wide body of knowledge regarding features in Northern Plains villages derived from previous excavations, coring, historical documents, and geophysical surveys. Importantly, a coring program performed at Huff Village imme-

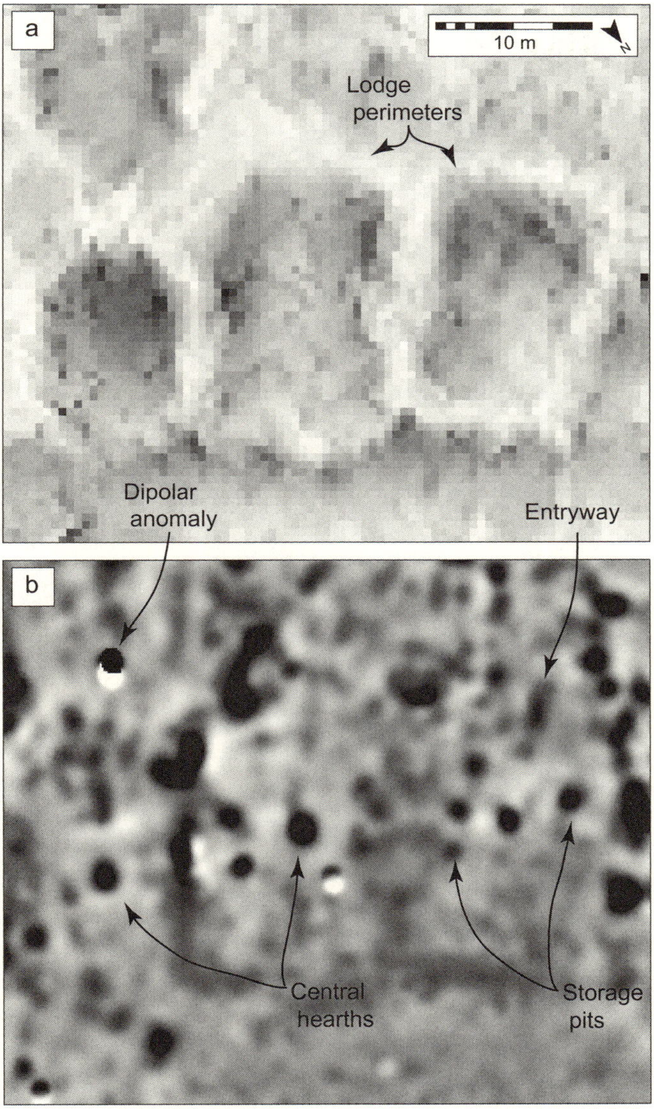


Figure 6.3. Close-up views of houses showing their perimeters and interior features in (a) sky view factor elevation data and (b) low-pass filtered magnetic data.

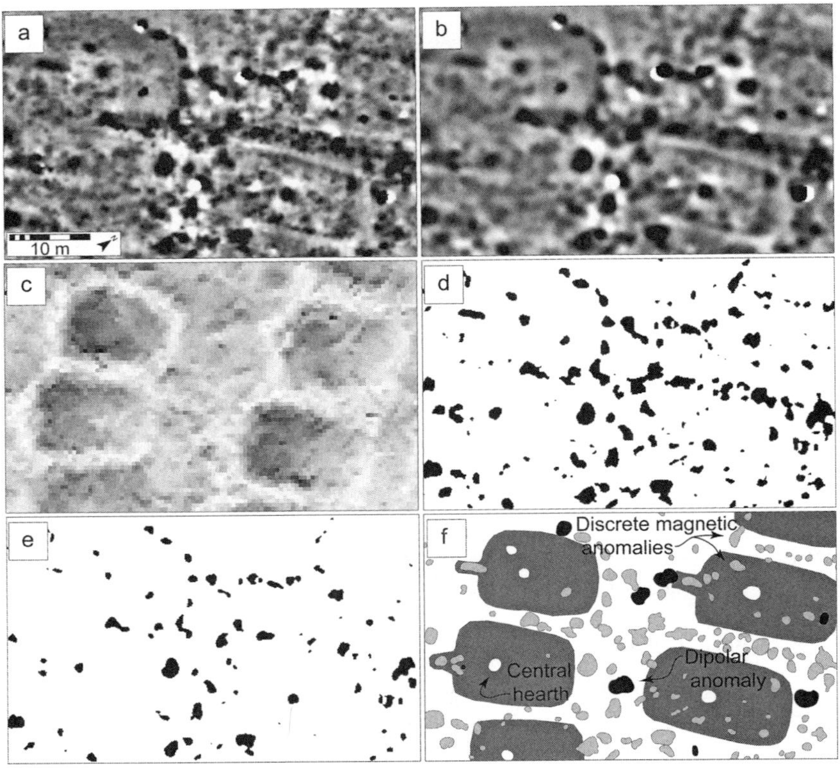


Figure 6.4. Close-up views, all to the same scale, of houses showing the (a) magnetic gradiometry data, (b) low-pass filtered magnetic data, (c) sky view factor elevation data, (d) 3 nT threshold layer, (e) 5 nT threshold layer, and (f) digitized features.

diately after the magnetic surveys in 1999 validated feature identifications and played a crucial role in my own interpretations.

A methodology for identifying discrete magnetic anomalies at Huff Village was adapted from that developed by Kvamme and Ahler (2000). This process involved the creation of several different layers in GIS (Figure 6.4). The full magnetic survey area exhibits thousands of anomalies, many blending together, causing identification of discrete anomalies and visualization of their patterning to be quite difficult (see Figure 6.1). For this reason, a circular low-pass filter 1 m in diameter was applied to the magnetic data set (see Figure 6.4b). The end result is a smoothed magnetic map, where much of the subtle magnetic noise is reduced, and discrete, roughly circular anomalies are made even more apparent. Additionally, threshold maps were employed to produce maps of magnetism above a certain level. In this case, threshold values of 3 nT and 5 nT were chosen, effectively eliminating magnetic noise less than those values and better approximating the sizes and shapes of magnetic anomalies (see Figure 6.4d, e). Comparison of the two threshold maps frequently revealed multiple sources (e.g., side-by-side storage pits) for large, irregularly shaped magnetic anomalies. Combined, these magnetic maps were used to identify and digitize magnetic features of likely archaeological significance greater than 0.25 $m^2$ in area (see Figure 6.4f).

## Results

Complete interpretation of these data sets yields a map of the village that differs considerably from that produced by Wood (Figure 6.5; compare with Wood [1967:Map 4]). The difference in the number of houses is perhaps not immediately apparent. Besides the 103 dwellings Wood (1967) mapped, including those lost to erosion and excavation, the magnetic gradiometry and elevation data reveal 12 additional houses. These new dwellings bring the total to 115 (a 10 percent increase) at a density of 23.8 houses per ha. Additionally, as many as 15 houses appear to be circular (or perhaps rounded square) rather than rectangular (Kvamme et al. 2009). Interestingly, Wood's excavated houses are on average 115.9 $m^2$ ($s$ = 36.5 $m^2$; $n$ = 10; note that Wood's [1967:31, 51] own dimensions indicate an area of 100.4 $m^2$ when an equation for the area of a trapezoid [$A = ((a + b)/2) * h$] is used to calculate the area for long, rectangular houses, although this estimate does not account for entryways), while the area of the remaining, unexcavated houses is considerably larger ( = 129.0 $m^2$; $s$ = 32.0 $m^2$; $n$ = 94). However, a Mann-Whitney test indicates that the difference between the two means is not significant ($U$ = 576; $z$ = 1.16; $p$ = 0.2447). The difference may indicate that my effort to digitize unexcavated houses, which lack posthole outlines like those visible in Wood's interpretive map of excavated structures, overestimates their average size. Yet lodges adjacent to the river's edge, the location of Wood's excavated dwellings, may actually be somewhat smaller, especially considering what appear to be much larger lodges encircling the plaza. In fact, Pearson's $r$ indicates a moderate negative correlation ($r$ = -0.59) between house size and distance from the plaza center for unexcavated houses (or $r$ = -0.62 if Wood's excavated houses are included). In other words, dwellings near the plaza tend to be larger than those farther from the plaza, perhaps signifying status differences among households at Huff Village. The smallest lodge is only 55.8 $m^2$, while the largest (excluding the single ceremonial structure) is 189.4 $m^2$. If the validity of area estimates in both data sets is accepted, the

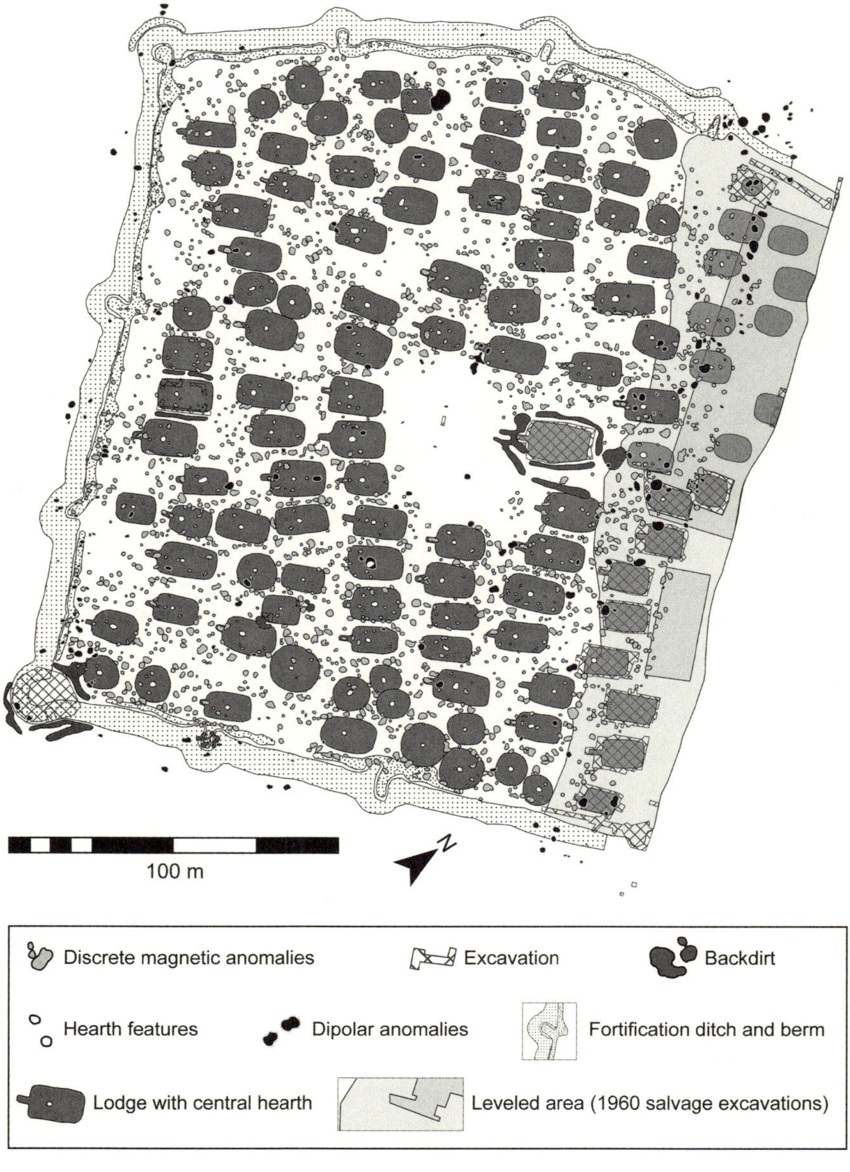


Figure 6.5. Interpretive map of Huff Village. Note the immense number of discrete magnetic anomalies, many indicating subterranean storage pits.

combined mean house size is 127.7 $m^2$ ($s$ = 32.5 $m^2$; $n$ = 104; excluding the dimensions for 11 houses because they are unknown). House size becomes a relevant statistic below.

Wood's (1967:31–32) excavations at Huff Village revealed a regular pattern for features within houses. For instance, he noted that most houses contained a primary fireplace along the centerline of the structure, with possible adjacent auxiliary hearths. Additionally, storage pits were located along the walls and in the corners of most lodges. Similar patterning is evident in the interpretive map, which shows a total of 2,311 discrete magnetic anomalies (excluding dipolar anomalies) greater than 0.25 $m^2$ in area in the space that has been unaffected by erosion or complete excavation (see Figure 6.5). If the area of the plaza (approximately 0.18 ha) and the village fortification are excluded, the density of discrete magnetic anomalies across the unexcavated portion of the village (approximately 38,526 $m^2$) is 6.0 per 100 $m^2$, an overall feature density lower than that estimated by Kvamme and Ahler from their limited sample. Based on this figure, it is estimated that another 234 features would have been located in the excavated portion of the site and the area lost to erosion (approximately 0.39 ha), for a total of 2,545 (note that the portion of the site lost to erosion prior to Wood's work is unknown, although some houses were truncated by the cut bank). Given the results of the previous intensive coring program at the site, which showed that 74 percent of cored features were storage pits, 14 percent were hearths, and 6 percent were middens, this would indicate perhaps as many as 1,883 storage pits, 356 hearths, and 153 middens. Magnetic features indicative of primary hearths are evident in each of the mapped houses at Huff, with the exception of houses impacted by erosion and excavation, and account for 101 of the anomalies (two are found in a single house), leaving approximately 250 auxiliary hearths within and outside houses. Thus, the overwhelming majority of the discrete anomalies represent storage pits.

## Discussion

Examination of these features reveals some interesting patterns (Kvamme 2007a:219; Kvamme et al. 2009:33–34). For one, storage pits are common around the exteriors of most houses and are particularly evident between adjacent structures (see Figure 6.4f). Additionally, many house entryways contain an elongated storage pit (confirmed during the coring program) (see Figure 6.3b). Discrete clusters of storage pits are common in locations away from houses, including lines of pits paralleling the fortification ditch. Finally, there is no appreciable difference between the number of storage pits excavated by Wood in each house and digitized storage pits identified within houses.

Based on the average volume of excavated storage pits (1.03 m$^3$) and the estimated number of pits at Huff Village, the estimated total storage capacity is nearly 1,939.5 m$^3$ (approximately 55,038 bushels). This value represents a significant difference from the total storage capacity of nearly 2,500 m$^3$ estimated by Kvamme and Ahler (2000) based on their limited geophysical survey of the village. Still, the number is impressive and attests to the great deal of time and labor that village occupants dedicated to agricultural production and the construction of storage facilities. Moreover, the storage pit estimate is conservative since many weakly magnetic features likely result from additional storage pits, and closely adjacent pits may often be identified as a single storage pit.

As noted by Mitchell (2013:81–84), historical data concerning agricultural yields among Plains groups are problematic because most postdate 1850, after which some fields were cultivated for the Indian agencies, the federal government provided support, new crops were introduced, and production methods were changing (e.g., see Will and Hyde's [1917] discussion regarding yields and these other factors). However, it is reasonable to assume that Plains Village farmers regularly produced between 20 and 30 bushels of corn per acre (Mitchell 2013:81–84). Given these yields, approximately 743–1,114 ha (1,835–2,752 ac) would have been necessary to account for the estimated total storage capacity at Huff Village (ignoring food that was consumed during or soon after harvest), a figure that may initially seem unreasonable but is only somewhat higher than nineteenth-century field sizes reported by Will and Hyde (1917). Clearly, the total storage pit capacity is a cumulative estimate of pits constructed over the lifetime of the village. Estimating the population of Huff Village yields additional insight.

Most methods for estimating the population of settlements using archaeological data rely on either total floor space or site size. Naroll's (1962) average of 10 m$^2$ of floor space per person and S. Cook and Heizer's (1968) 1.86 m$^2$ of floor space per person up to six and then 9.29 m$^2$ of space per person thereafter are commonly cited. With a house count of 115 and the average lodge size of 127.7 m$^2$, Naroll's method yields an estimate of approximately 1,470 people, while Cook and Heizer's approach suggests more than 18 persons per lodge. Alternatively, Wedel (1979) used historical data from Central Plains village groups to argue that 5.0 m$^2$ of floor space per person was more appropriate. This figure yields an unreasonably high estimate of about 25 persons per dwelling, however, and a total population of about 2,940. Other factors such as family status may also have influenced house size, a fact that led Roberts (1977) to produce a regression equation based on historically documented Mandan populations in which this is known to be true. His zero-intercept equation for Mandan populations yields an estimate of approximately 1,470 individuals at Huff Village, a number similar to that derived from Naroll's method and one that is supported by other historical information (Mitchell 2013:63–64).

This population likely resided in the village for only a short period, perhaps less than a decade, considering the lack of evidence of long-term occupation and data concerning lodge use life (Roper and Pauls 2005). The brevity of the occupation is supported by other archaeological evidence, including limited midden accumulation and few (if any) superimposed houses or house floors, rebuilt houses or house walls, or truncated pit features (Kvamme et al. 2009; Wood 1967). Given the population estimate of approximately 1,470 people, the per capita storage capacity for the settlement would have been around 1.3 $m^3$ (or 37 bushels). Of course, this estimate assumes that all storage pits were used simultaneously, which is unlikely given historical accounts of pit failure and abandonment, although they were reportedly used for multiple years (Will and Hyde 1917; G. L. Wilson 1917). Will and Hyde (1917) relate that gardens were typically 0.4 to 1.2 ha (1–3 ac) in size and that each mature woman had ownership of a garden so that approximately 0.1 to 0.4 ha (0.3–1 ac) was cultivated per person in each group. Only at the upper end of historically documented garden sizes and experimentally tested yields can the total and per capita storage capacity estimates be accounted for given the estimated population at Huff Village (Munson-Scullin and Scullin 2005; Will and Hyde 1917). Thus, the total storage capacity at Huff could reflect a year with especially high yields, although the annual total and per capita storage were likely somewhat lower given variability related to weather, insects, and soil conditions.

## Conclusion

This chapter presents one method for addressing site-scale anthropological questions using limited archaeological data as well as the results from recent extensive geophysical surveys. This approach, although imperfect, is especially important given the current era in which wide-area excavations are uncommon. Within these constraints, geophysical surveys can be used as much more than a tool to discover archaeological features. Although we must still base our ideas about changes in agricultural productivity in the Northern Plains during the fifteenth through the eighteenth centuries on limited excavation data, this work offers at least a baseline from which we may compare information from sites preceding and postdating Huff Village. Such changes in agricultural productivity have implications for larger issues such as developments in agricultural technology, the effects of climate on subsistence, and the influence of European colonization. While these larger issues are of interest to many, we often lack the empirical evidence necessary to address them. Geophysical data provide one means of establishing such empirically based understanding.

## Acknowledgments

Funding for the 2009 remote sensing investigations at Huff Village was provided by the State Historical Society of North Dakota. Fern E. Swenson, director of the Historic Preservation Division of the State Historical Society, and the entire staff provided considerable support and assistance in the field. The necessary instruments, computing equipment, and software were made available through the Center for Advanced Spatial Technologies and the Archeo-Imaging Lab of the University of Arkansas. Special thanks go to Ken Kvamme, the principal investigator, and Duncan McKinnon, who assisted with surveys. Finally, I thank W. Raymond Wood for the great deal of work he has undertaken at sites across the Plains.

# 7

# Transformation in Technologies

## A Look at Basketmaker III Archaeology in Southwestern Colorado

*Shanna Diederichs, Margaret Watters, Duncan P. McKinnon, and Bryan S. Haley*

The seventh century A.D. was a period of advancing technologies, migration, and transformation for ancestral Pueblo populations in the central Mesa Verde region of southwest Colorado. This period represents the stage between the adoption of agriculture and the development of permanent, sedentary villages, or what is known as a Neolithic transition, a stage of development difficult to isolate in other parts of the world (Ortman et al. 2011). In Earl Morris's view, fundamental Pueblo material traits appeared during this time and the "virile and progressive ensemble" of ancestral Pueblo culture came together (Morris 1939:19, 43). During this time in the Southwest the bow and arrow replaced the atlatl, beans and cooking pottery were introduced, public architecture in the form of Great Kivas appeared, and pithouse communities became sedentary as a result of a shift from subsistence gathering to farming (Kohler et al. 2008).

Architecture holds unique information about the scale, relationship, and development of communities. Unfortunately, the remains of pithouse architecture in the Southwest are notoriously difficult to characterize from the surface. The reasons for this are twofold. First, pithouse architecture is inherently perishable. Built from locally available sediment and timber, these shallow earthen structures easily deteriorate, often leaving minimal or no visible evidence of their location on the surface. Second, pithouse settlements were often constructed on what became prime agricultural land, which was reoccupied and farmed during the first eight hundred years of Pueblo occupation (after A.D. 750) and then by modern farmers over the last century. The "invisibility" of pithouse architecture presents unique challenges on several fronts: in discerning the presence of long-term habitation sites from short-term activity areas, in comparing the scale and organization of long-term habitation sites, and in developing reliable population estimates. These difficulties are more easily addressed for most Ancestral

Pueblo sites after A.D. 1000, when stone masonry became the architectural norm (Fewkes 1999; Plog 2008). As such, the challenges associated with surface visibility tied to pithouse architecture have significantly impacted the ability to define the full suite of cultural nuances associated with early Pueblo pithouse traditions and village occupations.

Archaeologists have dealt with issues stemming from the invisibility of pithouse architecture in several ways. Initially, early archaeological investigations identified burned pithouses for excavation by the fire-hardened and fire-reddened chunks of construction mud or jacal visible on the surface (Morris 1939). This method is productive when investigating specific populations with a tradition of burning structures during site abandonment. However, cultural resource management (CRM) work throughout the Southwest since the 1960s has demonstrated that only a portion of pithouse structures in the central Mesa Verde region are burned (Wilshusen 1999). Today, the most common technique for locating pithouses at early Pueblo sites involves heavy mechanized equipment. Either the sites are stripped of the upper deposits with a mechanical earth grader or consecutive trenches are dug across the sites with a backhoe (Chuipka 2009). This method locates and maps large structures across a site but disproportionately damages the more ephemeral portions of the archaeological record. While using heavy equipment offers visibility of broad areas and large prominent features, this method often removes the occupational surface surrounding pithouses, churns and reorganizes midden deposits, and limits the investigation of more nuanced details of site occupations. Furthermore, large-scale excavations involving the movement of hundreds of cubic meters of dirt to expose large areas is counterproductive where preservation of particular sites is the long-term archaeological conservation goal.

To this end, a few geophysical surveys have been employed to investigate early Pueblo pithouse sites in the central Mesa Verde region. During the 1980s, feasibility studies were conducted using ground-penetrating radar (GPR) to detect pithouse structures (Conyers 2013; Weymouth 1986). In some cases, this method was effective, but GPR waves attenuate at shallow depths in the dense eolian silt that covers much of the region (Conyers 2013). Additional geophysical surveys have used a combination of magnetic gradient and resistivity surveys to map the location of pithouse structures (M. Charles 2012, 2013). These were productive research-based student-training events, yet overall comparative visibility and site structure were limited as a result of their small-scale coverage (M. Charles 2012, 2013). Finally, marginal attempts have been made to integrate geophysical imaging into CRM site mitigation, but while results have been productive the efforts are few and unsystematic (Diederichs 2014; Horn et al. 2003; see also chapter 4, this volume).

As demonstrated in this chapter, noninvasive near-surface remote sensing surveys are a preservation-oriented and economically feasible method to pursue broad-scale exploration of pithouse distribution within an early Pueblo pithouse village. While a combination of large-scale magnetic gradient and resistivity surveys has been successful at later Pueblo period sites (Duff and Ryan 1999), the efforts discussed in this chapter are the first to apply these methods on a large scale to a pithouse era Pueblo site. The integration of these methods into early Pueblo archaeological research projects offers the ability to target excavations to investigate subsurface features proximate to pithouses (fences and burning pit features), evaluate the deposition of in situ midden deposits, and develop a better understanding of the nuanced details of site occupations and social organization in the central Mesa Verde region.

## The Dillard Site

The Dillard site (5MT10647) is a Basketmaker III (ca. A.D. 500–725) early Pueblo pithouse village located in the Four Corners area of the southwestern United States (see Figure I.1). During the Basketmaker III period, the adoption of domesticated food production is widely acknowledged as a fundamental technological advance that dramatically increased the carrying capacity for human life in marginal local environments and significantly expanded the ecological niches in which human populations could thrive (Kohler and Reed 2011). These transitional adaptations, known collectively as the Neolithic transition (Kohler et al. 2008; Simmons 2007), occurred in numerous locations around the world during the Early Holocene period as human populations adopted domesticated food and more sedentary lifestyles (Price and Gebauer 1995). This period represents a time of transformation and advancing technologies in Pueblo prehistory with the adoption of dryland or direct precipitation farming. At the Dillard site, occupants relied on the presence of deep deposits (1.75 m) of eolian-derived silty loam known as Mesa Verde Loess, suitable for dryland farming (Arrhenius and Bonatti 1965).

Artifact analysis, archaeomagnetic dating, and radiocarbon dating place the Dillard site between A.D. 580 and 700 (Diederichs and Copeland 2013; Diederichs et al. 2014; Wilshusen 1999).

The location of the Dillard site is important, as natural landmarks such as the San Juan Mountains to the east, the Mesa Verde cuesta to the south, and the lone Ute Mountain to the west are easily visible. These features create a unique and distinct landscape for the site. This expansive viewshed encompasses the central portion of the Mesa Verde region, linking the Dillard site to important cultural and geographical features. A recent LiDAR survey demonstrates the accessibility of the Dillard site to known contemporaneous Basketmaker III habitation sites in the surrounding settlement (Watters and Diederichs 2013).

In 2011, the Crow Canyon Archaeological Center (CCAC) initially targeted the Dillard site for excavation because of its potential to shed light on the foundation of social organization in Pueblo culture that accompanied the Pueblo Neolithic transition. Pueblo populations began to aggregate in earnest during the seventh century, when public architecture, usually in the form of a Great Kiva, was constructed at sites in newly colonized territories, at existing sites experiencing an increase in population density, and at sites now aggregated into protovillages (Schachner et al. 2012).

Great Kivas are circular structures that measure more than 10 m in diameter and are large enough for community gatherings (Wilshusen 1999). They are probably the manifestations of social institutions that mitigated issues associated with the Pueblo Neolithic transition such as sedentism, land tenure, and the risk that comes with small-scale farming in a marginal environment. Great Kivas are often regarded as the quintessential Pueblo architecture in the central Mesa Verde region because they were constructed continually throughout the entire Pueblo occupation. Initial excavations at the Dillard site documented the presence of a Great Kiva that is, at present, the only seventh-century Great Kiva confirmed in southwest Colorado (Fetterman 1991; Ortman et al. 2011). By investigating the Dillard Great Kiva and the pithouses in the village proximate to the Great Kiva, mapping and identifying pithouses in the surrounding settlement, and evaluating the intrasite organization and use of space at the site, the CCAC hopes to understand how the Dillard Great Kiva functioned to socially and politically integrate this early Pueblo farming community during the broader cultural transition to a more sedentary farming existence in the central Mesa Verde region.

## Interpreting the Dillard Landscape Organization

During initial investigations, the CCAC presumed that few, if any, habitation structures existed at the site, based on the lack of surface evidence in the form of topographic depressions, burned jacal (adobe), or soil staining. To confirm or refute these initial presumptions, Bill Volf of the Natural Resources Conservation Service (NRCS) surveyed (pro bono publico) a small area (20 x 40 m) south of the Great Kiva with electrical resistivity, which revealed several anomalies interpreted as potential structures (Volf 2011). While these results were highly informative, the overall research project plan did not include funding or resources to support additional geophysical surveys. As a result, the identified anomalies were excavated, precluding any additional geophysical surveys in the south portion of the site. The anomalies were confirmed as a concentration of pithouse structures, demonstrating that the geophysical survey was a highly effective way to characterize "invisible" site features. With this realization, the CCAC began searching for funding and partners to evaluate the north half of the site before its scheduled excavation in 2012.

The CCAC partnered with *Time Team America* (*TTA*) and Fort Lewis College to help identify pithouse structures and associated intrasite features in the north half of the Dillard site. In cooperation with the CCAC's Native American Advisory Group (NAAG), the CCAC holds to a principle of conservation archaeology. As such, the CCAC thought it was imperative to excavate less than 3 percent of the Dillard site while still attempting to understand the site scale, the number of pithouses and associated features, and the intrasite organization of the village. To uphold these principles, *TTA* and Fort Lewis College geophysical imaging specialists surveyed the north half of the site over four days of filming for an episode of *TTA* in May 2012 (M. Charles 2012; Watters 2012; Watters and Diederichs 2013). Methods included magnetic gradient (Kvamme 2006b), conductivity (Clay 2006), magnetic susceptibility (Dalan 2006), and electrical resistance (Somers 2006). This multimethod approach proved to be a powerful tool for creating an accurate plan map of the cultural and natural subsurface features, enhancing the ability to target excavations to investigate anomalies interpreted as pithouses and subsurface features surrounding pithouses, and providing a more accurate interpretation of the site sociospatial organization across the landscape. Geophysical and excavation goals at the site were to try to estimate site population, gain insight into site organization, and classify subsurface features to understand site context with the broader cultural and natural surrounding landscape.

Geophysical survey grids were established by CCAC surveyors and tied into the existing site coordinates. Because the inherent data collection rate differed for different geophysical survey methods, coverage varied with each method (see J. Johnson 2006). A magnetic gradient survey covered the entire targeted research area north of the Great Kiva to the site fence boundaries, whereas conductivity, magnetic susceptibility, and resistance surveys covered smaller areas. *TTA* geophysical imaging specialists conducted the magnetic gradient, conductivity, and magnetic susceptibility surveys, and the Center of Southwest Studies at Fort Lewis College conducted the resistance survey (M. Charles 2012). Magnetic susceptibility and conductivity data were not informative and are not discussed here (see Watters 2012 and Watters and Diederichs 2103 for a full report on coverage areas).

Ground-surface obstacles such as tall sagebrush and juniper along with surface rocks impacted geophysical surveys and subsequent data interpretation. Additionally, two areas of the site were recently burned during private land development, creating open "lanes" or windrows across the site. Finally, a few buried metal stakes created wide "halo" magnetic anomalies, essentially masking subsurface features in some areas. While these obstacles impacted final geophysical survey results (with numerous "no data" locations), they did not seriously impact the quality of final data maps or the visibility and interpretability of intrasite subsurface distributions (Kvamme 2008b). Each geophysical survey method contributed valuable information to the interpretation of the site, with magnetic gradient survey results providing the most in-depth coverage and extensive mapping of the subsurface features. As such, results presented in this chapter focus primarily on magnetic gradient data. The resulting interpretations from this survey method have guided subsequent excavations by CCAC archaeologists and confirm the presence of numerous subsurface cultural features related to the Dillard site occupation. Electrical resistance data are included in the discussion where corroborated with magnetic gradient data. Results significantly contribute to a better understanding of the history of the Dillard site as it relates to population estimates, social and community organization, and the use of space.

## Geophysical Results

Over the course of four days, a total of 0.96 ha (2.37 ac) of magnetic gradient data (Figure 7.1 left) and 0.20 ha (0.49 ac) of electrical resistance data (see Figure 7.1 right) were collected across the site to the north of the Great Kiva. The primary information gained from the magnetic gradient and electrical resistance surveys is an understanding of the broad-scale distribution of anomalies associated with possible pithouse architecture. In addition to providing the location of pithouse architecture, magnetic gradient anomalies are also interpreted as the remains of numerous postholes likely associated with fence architecture and isolated burned pit features not visible on the surface.

Eight possible pithouse anomalies were interpreted based on inductive (general shape and context) and deductive (magnetic and resistivity value)

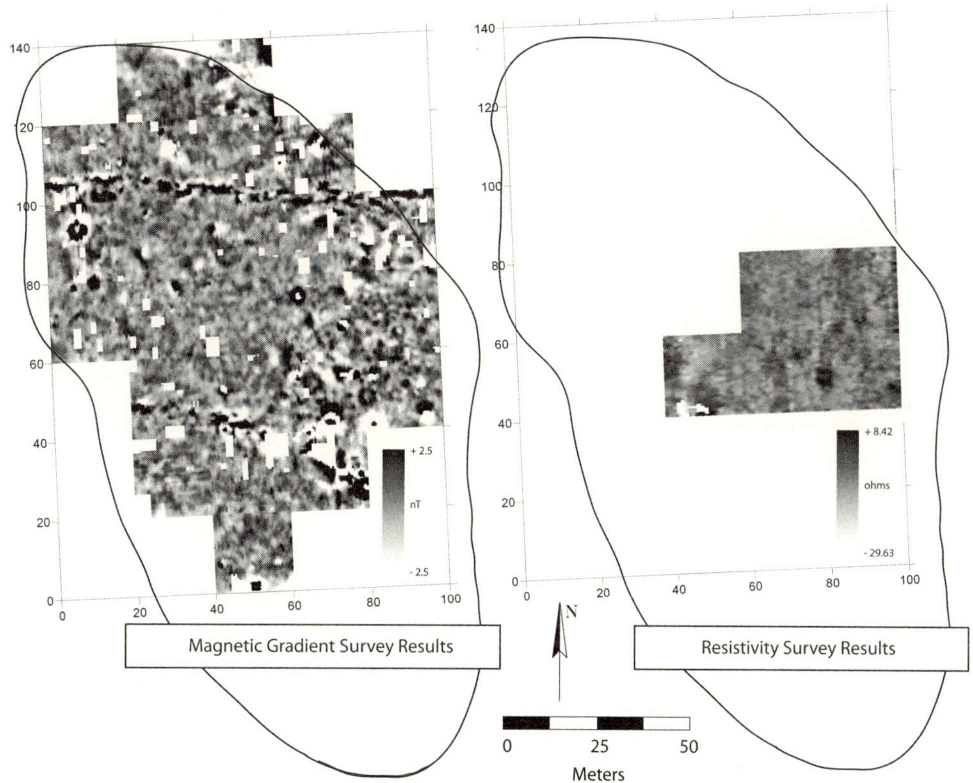


Figure 7.1. Magnetic gradient and resistivity survey coverage area. Magnetic gradient processing included a 2 x 1 low-pass filter biased in the *X* direction followed by linear interpolation at 25 x 25 cm. These final steps were used because the challenging terrain created some inevitable collection noise and artifacts.

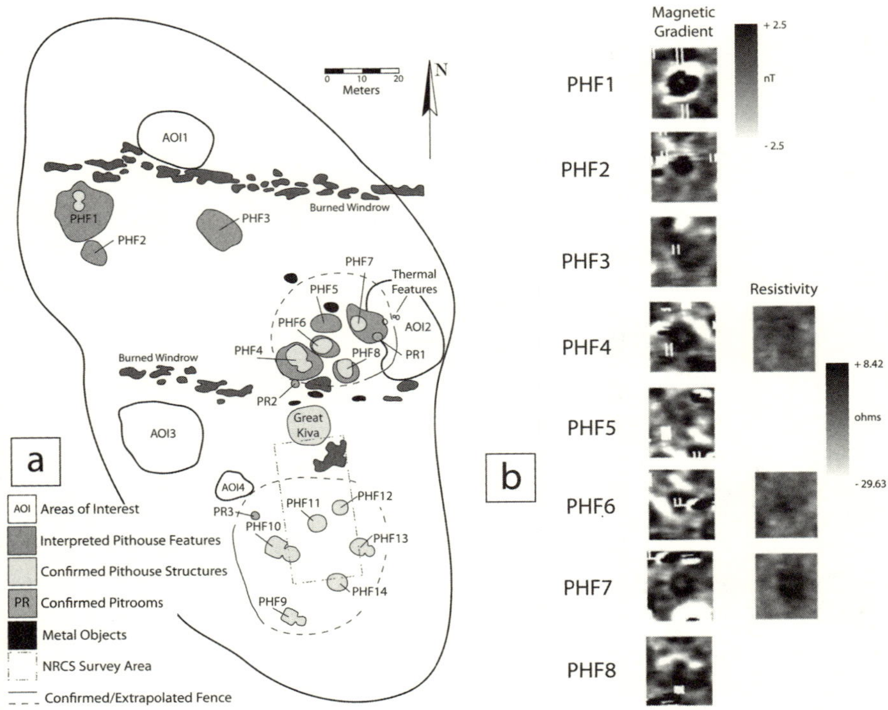


Figure 7.2. Interpretations with excavation results (a) and variability of pithouse anomalies in magnetic gradient and resistivity data (b). NRCS, Natural Resources Conservation Service; PHF, pithouse feature.

reasoning (Figure 7.2a, b). Five (PHF4–8) of the anomalies are located in a cluster directly north of the Great Kiva, and three (PHF1–3) are located about 40 m northwest of the Great Kiva. In addition to the identification of possible intact pithouse architecture, four areas were highlighted as areas of interest (AOI1–4) based on clusters of anomalies of possible cultural origin. In particular, magnetic gradient data documented a scatter of anomalies within a previously defined midden deposit downslope from the pithouse cluster north of the Great Kiva (see Figure 7.2a AOI2). The anomalies in AOI2 were concentrated in two parallel arcs encircling the possible pithouse cluster.

While the multimethod approach identified a total of eight possible pithouse anomalies, it is important to note the variability of interpreted pithouse features throughout the geophysical data set (see Figure 7.2b). While variability is present, confirmed pithouses (discussed below) were successfully imaged by electrical resistance, magnetic gradient, or a combination of both regardless of size, depth, shape, or configuration. The resistance results provide an important comparative data set to evaluate the presence of pithouse architecture, especially as it relates to differential burning.

## Excavation Results

The CCAC tested all eight possible pit-structure anomalies identified by geophysical imaging (see Figure 7.2a). To confirm the presence of a pithouse, each of the eight anomalies was tested to a depth of 1.5 m with a soil auger. In some cases, obvious roofing material in the form of burned soil and charcoal was identified, which corroborates the identification of high-amplitude magnetic anomalies (see Kvamme 2008b). Where geophysical images did not produce clear, unambiguous outlines of pithouse shape, size, or orientation, the anomaly was probed every 25 cm across both the north-south and east-west axes to better define the dimensions and extent of the structure.

Augering and excavation determined that three of the eight pithouse-sized anomalies were unrelated to subsurface structures (see Figure 7.2a PHF2, PHF3, and PHF5). Extensive augering of two of the anomalies (PHF2 and PHF3) found neither structural nor deposited cultural material. However, the soil extracted with the probes did not exactly match the highly developed red loess or caliche deposit recognized as the culturally sterile sediment at the site. Instead, sediments appeared to be a mottled mix of the recognized sterile sediments, suggesting geomorphic cut-and-fill episodes in these locations prior to site occupation. The third anomaly (PHF5) was tested with a 3 x 1 m excavation unit. The cultural deposits in this location were only 30 cm thick and consisted of moderately dense midden refuse and some discarded burned rock unrelated to pithouse construction. This deep cultural deposit is located just inside a possible fence line (discussed below), suggesting that it represents an accumulation of refuse and sediment just inside the fence.

The five confirmed pithouse structures (see Figure 7.2a PHF1, PHF4, PHF6, PHF7, and PHF8) were ultimately tested with 3 x 1 m excavation units to confirm the presence of a structure and to gather datable material. In the case of PHF4, a full half structure was excavated to more fully understand the pithouse construction details. Excavations confirmed that pithouse structures range in size from 2.5 to 6 m in diameter and from 0.20 to 1.75 m deep. The shape and spatial configurations of the pithouse features vary; some structures are fairly circular while others are subrectangular. The variations in excavated structures corroborate the variations identified in geophysical anomalies (see Figure 7.2b). At least three of the pithouse structures (PHF1, PHF4, and PHF8) contain antechambers (separate rooms), whereas others are single-room structures.

Outside and between pithouses, areas with a high density of small magnetic gradient anomalies were targeted for excavation, revealing extramural features and activity areas. Forty-eight square meters of extramural area were excavated in areas overlapping geophysical surveys in the north half of the site (AOI1 and 2). Earlier random 1 x 1 m testing in AOI2 identified three postholes below an identified diffuse refuse deposit. Subsequent to geophysical survey, four additional postholes and three shallow thermal pits were documented in an arcing alignment. While the posthole and shallow thermal pits confirm portions of a fence feature, extrapolation using point anomalies in the geophysical data appear to document the remnants of a loosely spaced fence surrounding the pithouse cluster to the north of the Great Kiva (see Figure 7.2a).

## Discussion

The geophysical methods used at the Dillard site were effective at mapping numerous pithouse structures and associated features related to the early Pueblo pithouse village. Additionally, the survey work has provided important contextual information for new insights into the organization of space and use of the village by the Dillard occupants. After four years of targeting geophysical anomalies with excavation, we now know that there are at least 11 pithouses and 3 pit rooms at the Dillard site and that these structures were used between A.D. 580 and 700 (Diederichs et al. 2014; Lengyel 2013). The largest structures are double-chambered pithouses with rooms measuring up to 6 m in diameter (Figure 7.3). The smallest are pit rooms, presumably used for storage, measuring about 1 m in diameter. All the buildings are semisubterranean, but their foundations range from 0.25 to 1.25 m deep. Each building was roofed with a superstructure of wooden beams covered by sediment. Evidence demonstrates that two of the pithouses (PHF 7 and PHF13) and the Great Kiva were burned completely at the end of their use life, while the other identified structures show signs of light burning during abandonment. All the structures at the Dillard site appear to have been systematically cleaned out when abandoned, which suggests that the burning of each structure was associated with the final closing of the building and was not just part of a series of accidental conflagrations.

### Population Estimates

Using geophysical and subsequent excavation data, we are now able to make a preliminary statement about the distribution of structures and an esti-

Figure 7.3. Example of pithouse architecture from Structure 5/6 at site 5MT20065 (H. Robinson 2014) (courtesy of Woods Canyon Archaeological Consultants, Inc., 1995).

mate of the population of the site. Absolute dates from the pithouses suggest that the site was occupied for less than three generations, mostly in the early to mid-600s (Diederichs and Copeland 2013; Diederichs et al. 2014). Widely accepted estimates for pithouse population (six to eight people per pithouse) derived from over 30 ethnographic comparisons (Lightfoot 1994; Varien and Potter 2008) suggest that the population of the Dillard site could have reached 50 people at its maximum. Importantly, the use of a broad-scale geophysical survey to document a more complete picture of cultural occupation in the north half of the Dillard site has offered more specific insights into the quantity and distribution of features (pithouses, fences, thermal features) and higher confidence in establishing population estimates.

Figure 7.4. Artist reconstruction of the Dillard site based on archaeological data including, but not limited to, structure footprints, roofing configurations, construction materials, and ground-truthed external features.

Unfortunately, because of the limited geophysical survey coverage in the south half of the site, we are unable to definitively reconstruct this area. The maximum population estimate for the Dillard site remains tentative.

## Site Organization

In addition to locating and illuminating pithouse structures throughout the village, the broad-scale geophysical approach allowed for the conceptualization of the spatial organization of structures across the landscape (Figure 7.4; Diederichs et al. 2014). For example, along the eastern edge of the site, magnetic gradient results identified a parallel pattern of small-scale single-point anomalies encircling the pithouse cluster north of the Great Kiva (see Figure 7.2a). None of these anomalies appear on the electrical resistance survey, likely related to differences in data sampling densities and pit fill material. Excavation of the eastern section of these linear anomalies identified the presence of widely spaced postholes related to a fence (see Figure 7.2a). Just outside the fence boundary are several thermal pits and small concentrations of fire-cracked rock associated with roasting (see Figure 7.2a thermal features).

Low-lying fences, referred to as stockades in the Mesa Verde region, are often found encircling a household at single-habitation Basketmaker III sites (Wilshusen 1999). The evidence for a fence at the Dillard site encircles four households and their associated storage rooms. At present, this is the first evidence of a fence enclosing more than an extended family–sized Basketmaker III compound. With the geophysical and excavation data serving as a model to investigate the potential of a similar fence enclosure, additional excavation in the south half of the Dillard site revealed a second fence feature (see Figure 7.2a). These fences appear to effectively divide two household compounds into two specifically delineated groups. The broad-scale geophysical survey in concert with excavations reveals not only important information on site organization and the distribution of the more nuanced features but also begins to provide valuable insight into social and community organization and the use of space within the Dillard village.

## Conclusion

This work demonstrates that magnetic gradient surveys are the most efficient tool for mapping Basketmaker III sites in the region (e.g., similar archaeological features, ground matrix, and environment). However, because magnetic anomalies often extend past the boundary of the actual feature (see Kvamme 2006b), an electrical resistance survey produced accurate images of pithouse shape and size, allowing for more precise targeted excavation. This multimethod approach demonstrates that a large-scale magnetic gradient survey to map the broad landscape distribution, combined with an electrical resistance survey to highlight and better define particular anomalies, can effectively map a pithouse village and constituent features that are "invisible" from the site surface.

Based on results from the Dillard site, it is recommended that future site explorations within the central Mesa Verde region incorporate geophysical surveys as a preservation-oriented and economically feasible method to pursue the broad-scale exploration of pithouse organization within Pueblo pithouse villages. This approach enables identification of pithouse organization and enables us to target excavations to investigate subsurface features surrounding pithouses, evaluate deposition of in situ midden deposits, and develop a better understanding of the nuanced details associated with site occupations and cultural traditions. The geophysical survey along with excavations at the Dillard site has been valuable in addressing site settlement questions related to population estimates, providing insight into the organization of the site and identification of both pithouses and more ephemeral features. It has also helped create an interpretive framework associated with the use of space at the site and the overall relationship of the site within the broader cultural and natural surrounding landscape. With the geophysical investigations at the Dillard site providing unanticipated new information for interpretation, we are able to look at the site as a whole and introduce new and more nuanced contextual interpretations to the investigation of the Basketmaker III people who lived at the Dillard site. Certainly, these revelations will drive future research questions at the Dillard site and in the field of Basketmaker III research in the central Mesa Verde region.

## Acknowledgments

This research was conducted with contributions from the Crow Canyon Archaeological Center archaeologists, interns, research associates, and support staff under the direction of Shirley Powell, and students of Fort Lewis College under the direction of Mona Charles. Bruce Barrow, director of the second season of *Time Team America*, was instrumental in developing the project. Finally, our work could not have been completed without the support and hospitality of Jane Dillard. For this we are sincerely thankful.

This work was undertaken as part of the filming of season 2 of the PBS prime-time program *Time Team America*. The program is coproduced by Oregon Public Broadcasting and Videotext LLC and funded through a grant from the Colorado State Historical Fund and a National Science Foundation Informal Science Education grant (no. 1114113). Any opinions, findings, conclusions, or recommendations expressed in this material are those of the authors and do not necessarily reflect the views of the Colorado State Historical Fund or the National Science Foundation.

# 3

# Archaeological Landscapes

# 8
# Conceptualizing the Anthropogenic Islands of Southern Florida with LiDAR

*Victor D. Thompson*

The analysis of space and architecture in the ancient world presents a number of theoretical and methodological difficulties for archaeologists. Scholars have made great strides in understanding the organization of communities from small-scale hunter-gatherer sites to large preindustrial urban centers (e.g., Kroll and Price 1991; M. Smith 2007; V. Thompson and Pluckhahn 2012). At every scale of human settlement, we can find intentional actions and the ordering of space. Our understanding of such site layouts relates to a host of social factors including domestic space, mobility, large-scale labor projects, monumentality, social inequality, and city planning, among others (M. Smith 2007). We face two primary difficulties when interpreting any of the aforementioned social issues. First, we must have an appropriate understanding of the scale and extent of the material evidence related to a specific phenomenon (e.g., the distribution of high-status houses compared with low-status ones). The second, related issue is typological. While typologies are sometimes useful if done well, they can also obscure important variation if constructed poorly. In worst-case scenarios, such typological systems are simply inappropriate for a given area of study. Certainly, these two issues are familiar to other areas of archaeological inquiry and are discussed most often in the context of ceramic analysis (see Trigger 2006).

Perhaps the most widely accepted typological systems in the Americas are those used to describe extant architecture. Although there are regional variants, a number of common classes of architecture often, in turn, play into site hierarchies and other intersite classification schemas (e.g., village site, camp, civic-ceremonial center, mound center). In particular, for archaeologists working in Mesoamerica and the American Southeast, the terms "mound," "plaza," and "village" are frequently used to describe much of the architecture of these areas. This seemingly unproblematic vocabulary has been in use in these regions for over a century (e.g., Cushing 1896).

"Mound," "village," "domestic area," and the like can be useful terms, which is why these classifications have such a long history in archaeological practice. However, new methods, as well as improvements upon older technologies, are beginning to indicate that such classifications are more problematic, at least for some areas, than previously thought and carry with them connotations that obscure insight into social processes or the behaviors that created such features (see Fish et al. 2007 for some of these issues related to Trincheras sites in the American Southwest). Kidder's (2004) research on plazas in the lower Mississippi Valley underscores this point nicely. He argues for plazas to be considered architecture on the basis that they were "specifically engineered in the same way that mounds were deliberately built" (2004:529). This is a far cry from plazas being defined simply as empty space surrounded by mounds. Similarly, Sherwood and Kidder's (2011) work on mound construction also emphasizes the complex processes that are obscured by the relatively simple term "mound." They show that these constructions required a great degree of planned labor that included knowledge of engineering and considerable knowledge regarding the physical properties of soils. These two brief examples hint at the possibility that our commonly used terms for architecture are obscuring many of the processes and behaviors that make such ancient creations inherently interesting to archaeologists, as well as to the general public.

Recent advances in methodological and theoretical aspects of remote sensing are allowing archaeologists to know both the scale and extent of given phenomena (Conyers and Leckebusch 2010; Kvamme 2003a; V. Thompson et al. 2011). Such knowledge is critical to our ability to understand the underlying principles in site feature typologies. Airborne LiDAR (light detection and ranging), in particular, is one such technique that is now available. Airborne LiDAR allows the mapping of sites as well as entire landscapes at a scale and level of detail that were previously not feasible because of time constraints or environmental conditions (e.g., Chase et al. 2011; Devereux et al. 2005; K. Johnson and Ouimet 2014; Rowlands and Sarris 2007). In this chapter, I use LiDAR to image and analyze the large shell-bearing sites of southern Florida (see Figure I.1). The purpose of this is to explore the degree of organizational complexity among site layouts along the southwestern coast of Florida.

Through this analysis, I examine similarities of site layout. In exploring these issues, I further the argument that the Native Americans of this region altered the landscape on a scale previously unknown for groups following a fisher-gatherer-hunter economy (see Schwadron 2010a:272, 2010b; V. Thompson et al. 2013:191; V. Thompson and Pluckhahn 2014). In addition, this analysis shows that these sites show evidence of at least the same degree of organization as sites occupied by contemporary societies reliant on agriculture in the interior Southeast.

## Anthropogenic Islands of Southern Florida

Shell-bearing sites in the American Southeast are prompting renewed attention in archaeological circles. At the center of this is an ongoing debate as to the fundamental nature of these sites. That is, do they represent monumental architecture and ritualized feasts, or are they merely the remains of quotidian meals and daily practice? (See V. Thompson and Worth 2011.) Such discussions, however, are not localized to the American Southeast and can be found in the literature on both Australia and coastal California sites (e.g., Luby et al. 2006; McNiven 2012). Most recently, Marquardt (2010a, 2010b) has challenged archaeologists working on shell-bearing sites to take a sediment-oriented approach and pay more attention to their environmental context. I also suggest that in addition to this microscale sedimentological approach, we take a quantitative approach to the overall form, layout, and physical properties (slope, volume, etc.) of these sites.

To illustrate the value of a macroscale approach, I examine a sample of sites in the Caloosahatchee and Ten Thousand Islands regions of southern Florida. Much of the matrix of these sites is shell and other anthropogenic sediments (Marquardt and Walker 2012, 2013; Schwadron 2010a), and in some cases these sediments form whole islands. The various features of these sites go by various names including mound, platform mound, temple mound, ridge, shell work, shell ring, water court, shell field, flat area, and water feature, among others (see J. W. Griffin 2002:274–275; Marquardt 2010a, 2010b; Schwadron 2010a, 2010b; Widmer 1988). Based on the form and layout of such sites, some of these categories seem to be somewhat arbitrary, inhibiting discourse between scholars.

While not individually critiquing the terms listed above, I think the issue of platform and temple mounds provides an illustration of the systemic issues at play in using such terms for these shell-bearing sites in Florida. Marquardt (2010b:562, 2014) notes that some researchers use the term "temple mound" to refer to certain mounds in southwestern Florida (see Widmer 1996:24). Marquardt (2010b:563) takes issue with this designation, as no archaeological evidence necessarily suggests an exclusive temple function for the structural evidence on top of mounds. Indeed, Marquardt questions the validity of the term "mound" for many of the deposits identified, suggesting that the evidence provided (i.e., large deposits of so-called clean shell) does not necessarily indicate intentional mound construction. Here, Marquardt (2010b:566) is not saying that there are no mounds, but that the case for mound construction must be made with evidence, which in his view begins with a sedimentological approach. I add that we must also consider the temporality of deposition in such arguments, as this is key in evaluating how rapidly a deposit forms (V. Thompson and Andrus 2011).

In what follows, I offer a preliminary analysis of a selection of sites to begin to explore some of their commonalities and how they might differ from other settlements in the American Southeast. Such an analysis has already been attempted; however, earlier efforts used architectural typologies that may not be wholly appropriate for southern Florida. The second advantage that the present analysis has is the use of LiDAR-derived maps. Airborne LiDAR presents a more accurate picture of the landscape than conventional maps of these often densely forested sites in subtropical Florida. Conventional mapping in these environments would require thousands of hours of clearing in addition to mapping with a total station. Obviously, airborne LiDAR allows for mapping of these sites at a far greater resolution over a shorter period of time.

## Conceptualizing Layouts

Since 2009, I have been involved with a number of colleagues and students in the analysis of LiDAR from sites in Florida and along the Georgia coast. We have amassed a database of scores of sites, the analysis of which is ongoing (see Colvin 2014; Pluckhahn and Thompson 2012; V. Thompson et al. 2014). I focus on four of these sites (Mound Key, Dismal Key, Russel Key, and Fakahatchee Key) to demonstrate the utility of the approach outlined in this chapter and its implications for discussions of site layout and architectural features of the region.

All LiDAR used in this analysis is publicly available via the National Oceanic and Atmospheric Association website (http://coast.noaa.gov/digitalcoast/). All aerial LiDAR data were processed in ArcMap 10.2, with the LAS toolset in the program using the NAD 1983 and NAV 1988 horizontal and vertical datum, respectively (see Turck 2013 for step-by-step instructions). Although we have noted some issues with these publicly available data sets, these can be mediated either through the fusion of conventional mapping data or through preprocessing with third-party software, such as Fusion (Pluckhahn and Thompson 2012; Rochelo et al. 2015). The vast majority of coastal sites, especially those in this analysis, tend to have good coverage densities and correctly classified returns (e.g., ground, water). The total area and number of LiDAR data exchange file (LAS) points, including returns for ground and water, for each data set are as follows: Dismal Key (1,139,800 points; 1,423,842 $m^2$), Fakahatchee Key (994,942 points;

1,754,144 $m^2$), Russell Key (214,466 points, 804,000 $m^2$), Mound Key (202,364 points, 638,841 $m^2$).

Once digital elevation models (DEMs) of the sites were created, I followed much the same procedure that my colleague and I used for our analysis of the Crystal River site (Pluckhahn and Thompson 2014). In that analysis, our primary concern was how Crystal River's layout diverged from that of other sites in the region. In the present analysis, my concern is the similarities among these sites that help in understanding the nature of site layout and how conventional architectural terms may perhaps inhibit our understanding. As with the Crystal River study, I use M. Smith's (2007) principles of elaboration for ancient cities as concepts for this analysis. These concepts include axiality, orthogonality, and symmetry.

Axiality for this analysis is the use and number of avenues or straight lines incorporated into the site layout (M. Smith 2007:12). Because of the complexity of the site layouts, my analysis includes only a minimum number estimation, as there are likely more than are readily observable in the digital elevation models created from the LAS data sets. However, this provides a baseline of comparison across the various case studies.

Orthogonality, according to M. Smith (2007:12), is the use of right angles in a site layout. Smith alludes to several different processes that can result in such use of right angles. The first is what he refers to as "semiorthogonal urban blocks," which are the result of practical and efficient building practices—for example, as new houses are constructed they will often abut earlier structures at right angles (13–15). Such examples are in contrast to more formal uses of right angles, which Smith (15) refers to as an "integrated orthogonal plan" that uses right angles as part of the plan of public buildings, as in Teotihuacán's avenue of the dead. Still others may even be modular in structure with repeated, regular avenue layouts such as those found in Greek cities (16).

Symmetry is the last of M. Smith's (2007:12) concepts in the present analysis. For our Crystal River analysis we defined it as parallel axes, as well as symmetrical placement of architectural features (i.e., the burial mounds) (Pluckhahn and Thompson 2014:67). While neither M. Smith (2007) nor Pluckhahn and Thompson (2014) discuss types of symmetry, a number of layouts can be symmetrical, including the reflection and translational symmetries that I explore in this study.

## Identifying Patterns

Upon examination of the LiDAR images, the first thing that should be clear is that all these sites are extremely large by Southeast standards, and many

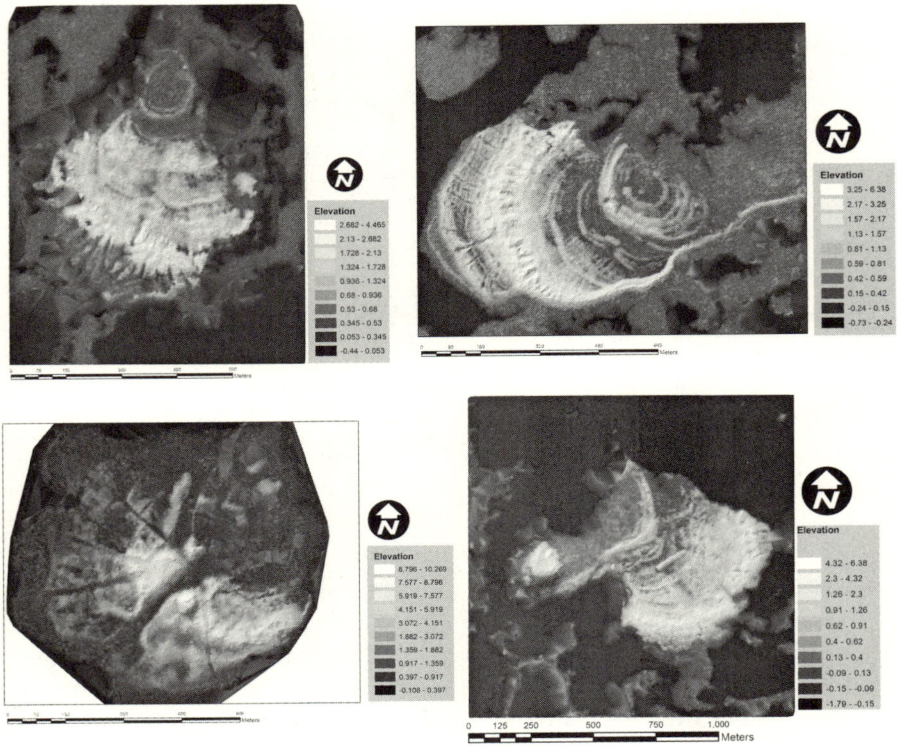


Figure 8.1. LiDAR image maps of four southern Florida sites: top left, Russell Key; top right, Dismal Key; bottom left, Mound Key; and bottom right, Fakahatchee Key.

dwarf Mississippian sites (Mound Key = 51 ha, Dismal Key = 30 ha, Russel Key = 23 ha, and Fakahatchee Key = 40 ha) (see Schwadron 2010a:134) (Figure 8.1). They also have tall landforms or "mounds," some of which exceed 10 m above modern sea level. Although Mound and Fakahatchee Keys have landforms that can be positively identified as mounds, in other portions of these sites, and in the other two cases in this study, it is not clear where one "mound" ends and another begins. Indeed, this has often led to some confusion regarding the labeling of features in the region. Finally, the other key point is that it is likely that a large portion of the sediments that make up these features and landforms were once consumed as food (see Marquardt and Walker 2013). This is not to say that all such sediments represent the simple disposal of refuse underfoot, although this certainly may be the case for some deposits. However, this is an important distinction from their earthen counterparts of the interior Southeast.

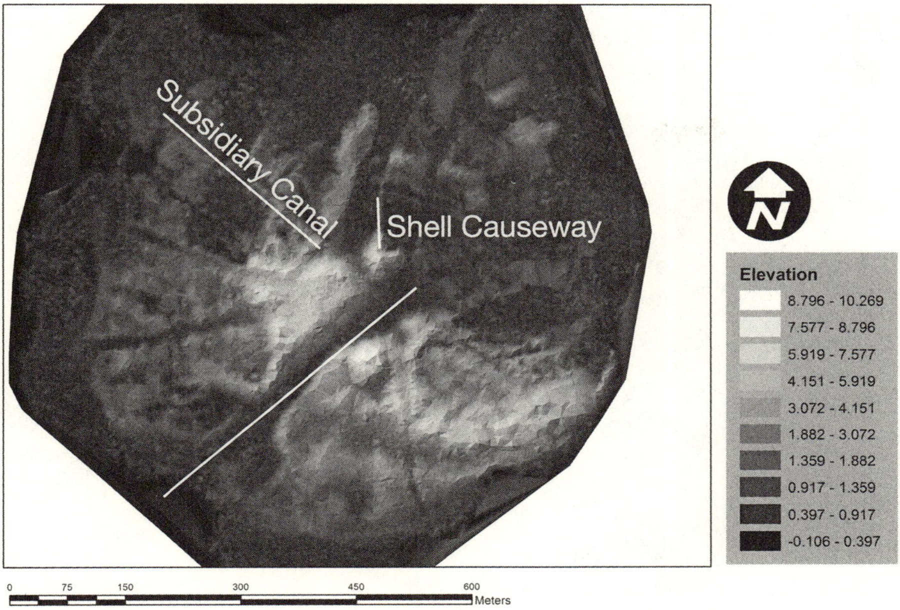


Figure 8.2. LiDAR image map of Mound Key showing the central main canal, linear shell causeways, and orthogonal subsidiary canal.

In terms of axiality, these site layouts include a number of straight-line avenues. Most interesting is that these avenues include raised causeways, as well as canals, for vessel transport. The most obvious examples of both of these come from Mound Key, which has a central canal that bisects the island (Figure 8.2). In addition, a raised shell causeway connects two landforms across an expanse of what is now covered with mangroves. Not all raised causeways are straight-line avenues, as can be seen in the map of Dismal Key, which has an undulating shell causeway leading up to the higher landforms along the western portion of the site (see Schwadron 2010a: 136–137).

Many of the straight-line causeways appear to be on the fringes of the island sites. These may in fact be related to either large stilt houses or water control features. Whatever the case, each of these sites has a large minimum number of these straight lines incorporated into the overall pattern, with likely many more not readily identifiable in the data ($n$ = 9 for Mound Key, $n$ = 31 for Dismal Key, $n$ = 20 for Russell Key, and $n$ = 27 for Fakahatchee Key). These data show that many features are straight avenues in excess of 20 m long.

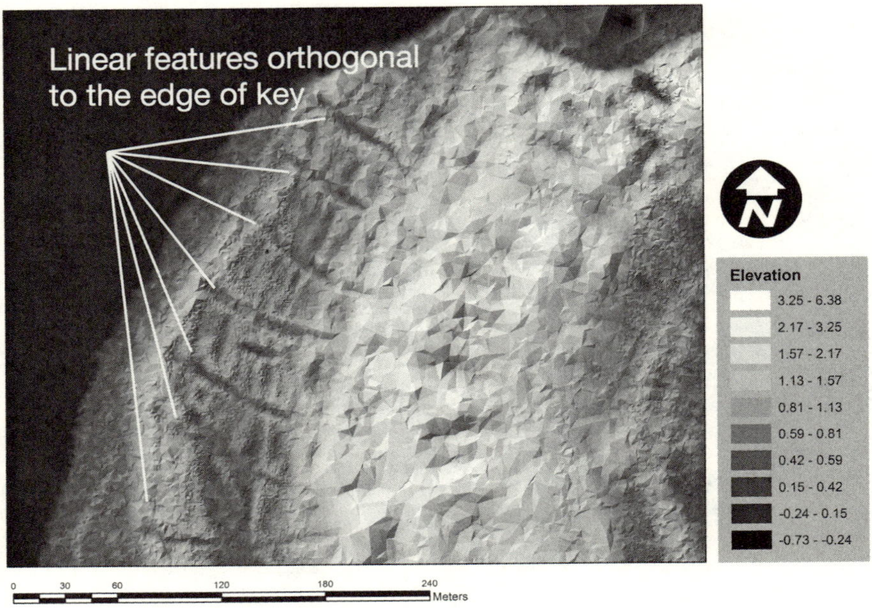


Figure 8.3. LiDAR image close-up map of the northwest section of Dismal Key showing the orthogonal pattern of shell ridges and troughs.

Although it is less apparent, these sites do appear to exhibit some degree of orthogonality. Many of these right angles can be found along the edges of the sites in semirepeating patterns. As we do not yet wholly understand the function of some of these features and how they are created, we cannot say that they represent formal orthogonal planning. In fact, it is likely that practicality, function, and disposal practices were the processes behind this pattern on the fringe areas (Figure 8.3). This most closely resembles what M. Smith (2007) calls semiorthogonal urban blocks, which is a special case. For these Florida sites I would simply call it a semiorthogonal pattern, given that the terms "urban" and "block" do not really apply.

In contrast to the semiorthogonal patterns already noted, the patterns in at least three of these sites, Mound Key, Russell Key, and Fakahatchee Key, exhibit right angles in terms of either canals or ridges that extend off the main canals at the site (see Figures 8.1 and 8.2). Russell Key and Mound Key may be the two best examples. Russell Key has a main canal and a central depressed area that bisect the island at right angles. Mound Key has a minor canal bisecting the main canal to the northwest and southeast. Indeed, Mound Key is likely the clearest example of what may be an integrated orthogonal plan for these sites. Less clear are the right angles off the main canals at the other sites. However, it is difficult to tell whether this is a true integrated pattern or a product of other factors (e.g., refuse deposition around houses). Smaller canals shooting off at various angles can be seen at other Calusa and related sites as well, with Pineland being the best studied and most intensively documented (Marquardt and Walker 2012, 2013; V. Thompson et al. 2014).

Many researchers note that there is some degree of symmetry in these sites, often suggesting that there is some underlying formal plan (Beriault et al. 2003:46; Marquardt 1992:47; Schwadron 2010a:59, 133; Torrence et al. 1994:13). Indeed, this is perhaps the most interesting aspect of the sites in this study. It seems that all of them exhibit symmetry to one degree or another. As with most archaeological problems, the degree to which this is recognizable depends on the scale at which the site layout is examined.

To an extent, the identification of bilateral symmetry is correct for all four of the sites in this study. Russell Key and Mound Key exhibit the most bilateral or reflection symmetry. Schwadron (2010a:136–137) notes this especially for Russell Key and states that the canal and central depression divide the key into four quadrants, with the southern two being a bit larger than their northern counterparts. As on Russell Key, the canals on Mound Key appear to divide the site into different sections, though the proportions of these sections differ in size. What is most interesting about this pattern is that the northern section, it seems, is the location of the known burial mounds.

In contrast to the above two sites, Dismal and Fakahatchee Keys do not appear to exhibit the same degree of reflection symmetry. As Schwadron (2010a:133) notes, both of these sites have similar layouts compared to other sites (see also Key Marco [Cushing 1896; Schwadron 2010a:152]). What makes these two different in this study is that the axis of symmetry shifts in both of these sites (Figure 8.4). Thus, only sections of these sites have a high degree of reflection symmetry. In addition, Fakahatchee Key has an offset mound with burial interments (Schwadron 2010a:127), which adds a certain degree of asymmetricality to the site.

In addition to reflection symmetry, all these sites exhibit a high degree of translational symmetry. Here "translation" means that "the same bound holds for the distance between any two distinct lattice points" (Stewart 2013:60). In the case of these sites, the so-called finger ridges that often represent the highest points of each of the four keys are the best example of this kind of symmetry (see Figures 8.1 and 8.4). In addition, Fakahatchee Key appears to have chambered water courts that exhibit some degree of both reflection symmetry as individual features, and translational symmetry in the apparently repeated patterns along the edges of the site (Figure 8.5).

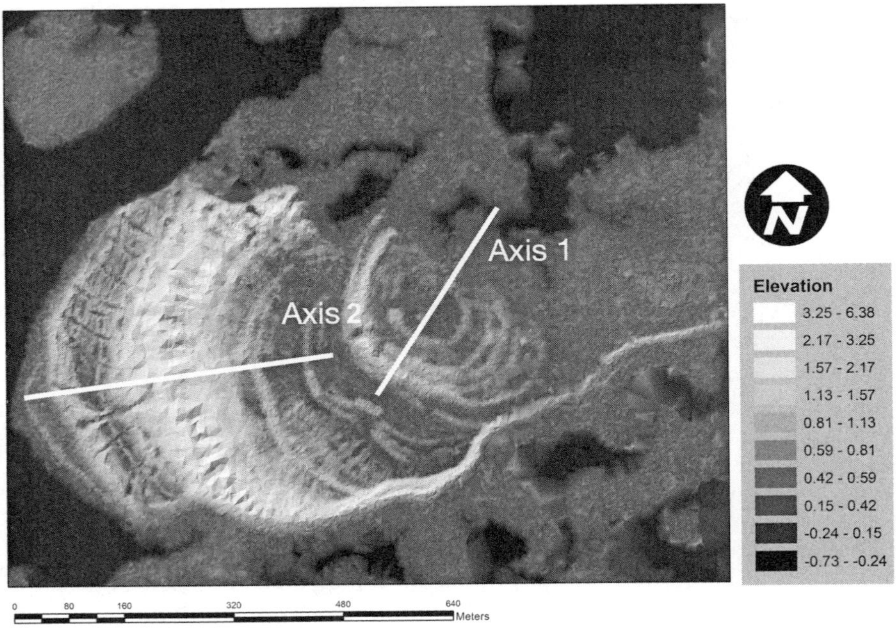


Figure 8.4. LiDAR image map of Dismal Key showing the shift in the axis of symmetry between axes 1 and 2.

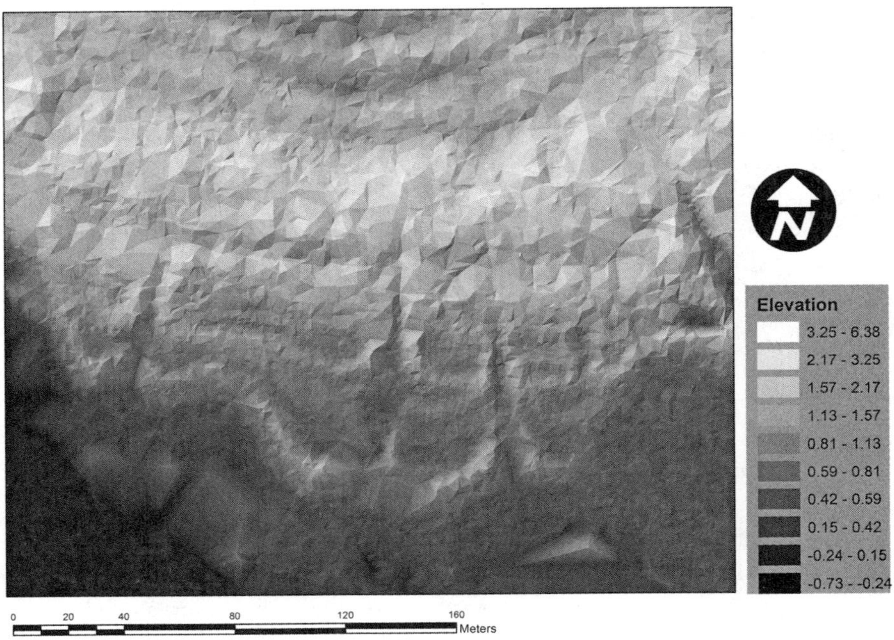


Figure 8.5. LiDAR image map of the southern portion of Fakahatchee Key showing translational symmetry in chambered shell features along the edge of the key.

## Discussion

Marquardt (2010b) rightly points out that "functional interpretations of shell works based solely on inference are less than convincing." Further, he cites Milner's (2004:306) important point that just because a site's features seem to be organizationally related it does not mean that there was a fixed plan from the beginning. Given that there appears to be evidence that the massive shell mounds of southern Florida were repurposed and moved around to some degree (V. Thompson and Marquardt 2015), what we may be seeing is the collective manipulation of a site or the implementation of design either over time or at the very end of a site's occupation. If we accept this then it frees us, at least for this study, from knowing whether there was a plan at the outset. Instead, because we do not know the chronological and developmental history of each landform, a more productive approach based on the currently available data is that we view site layouts as the product of the cumulative use of such places, although at some point a microhistorical and depositional approach to these sediments will shed additional light on organizational planning.

As Schwadron (2010a:54) states, most researchers have now concluded that the large shell-bearing sites of southern Florida are not simply refuse accumulations, but planned sites to some degree. Indeed, as she eloquently states, these sites are "purposefully constructed features that suggest architectural planning and landscape terra-forming to construct functional features, and to define public, domestic and ceremonial spaces" (61–62). Given the present analysis, I agree and believe that we can now begin to identify the elements of site layout in the region in a more formal manner that moves us beyond simply identifying planned and unplanned (see M. Smith 2007:4 for this point).

At least for the sites in this study, clearly the cardinal directions or any regular offset from one do not play a discernible role in the site layouts. At other Native American sites, archaeologists posit an important role for solar and lunar alignments and the use of cardinal directions in site layout (see Pluckhahn and Thompson 2009 for a discussion of this for Crystal River). This, however, just does not seem to be the case here. It may be that there is some other organizing principle that operates on a more local level (e.g., wind direction, tidal flow).

In all the sites in this study, there is at least some degree of axiality and orthogonal patterns. Some of the straight-line avenues include not only those that would have directed pedestrian traffic, as in the case of shell ridges, but also aquatic travel, as in the case of canals. In addition, some of these avenues of movement also have perpendicular causeways and canals that move off at right angles, suggesting that at least some degree of orthogonality played a role in site organization. In other areas of the sites, right angles emerged in a semiorthogonal fashion thanks to more pragmatic use of space, rather than incorporation into a grand plan.

What is most interesting regarding these sites is that the degree of axiality and orthogonality is much higher, or at least more obvious, than at other sites in Florida. For example, in the study of Crystal River some degree of both of these is present (Pluckhahn and Thompson 2014). However, instead of indicating true avenues, the placement of mounds and the right angles of the mounds themselves point to how these attributes were used. This is very different from the sites of the above studies, in which straight lines and orthogonal causeways would have structured physical movement around the site.

Given the sheer number of various landforms and features at these sites, it is perhaps a bit amazing the degree to which they exhibit various kinds of symmetry. Indeed, at other sites with fewer features, such as a few mounds arranged around a central plaza, it would be easy to point to the likely reflection symmetry in the site layout. While these sites do appear to have some reflection symmetry, it appears that it may not be expressed in some cases over the whole site. As I point out, in at least two cases it seems that the axis of symmetry changed throughout the life history of the site. Thus, in at least these cases, it appears that the organizational plan changed or perhaps some other localized environmental change occurred to alter the way in which people used and organized their space.

In addition to the reflection symmetry there appears to be translational symmetry in these sites, particularly along the fringes. In some cases this appears to be planned, such as in the chambered water courts (see Figures 8.4 and 8.5). In other cases, such as the addition of new "finger ridges" along the edges of these sites, this symmetry may simply be a result of population growth and the addition of new households or lineages. This perspective should not be thought of as detracting from the organizational complexity of the site layout. In fact, it might indicate a more complex overall plan, one that incorporated population growth yet still maintained regularity in use of space across the site as a whole.

Perhaps the most important point regarding the symmetry of these sites, at least for Mound Key and Russell Key, is that the axis of symmetry is not simply an imaginary line but an actual intentional construction in the form of a canal. This is perhaps the strongest line of evidence that these sites are planned to some degree. Canals among the Calusa and in southern Florida in general served a number of practical, economic, and ceremonial functions (see Luer 1989; Marquardt and Walker 2013). Thus, it is perhaps unsurprising that they were the central axis of symmetry. Other sites not included in this study also seem to follow this pattern (e.g., Big Mound Key, Pineland; see Marquardt and Walker 2013).

In summary, it seems that there are some regular elements outlined above that are incorporated into some southern Florida sites. Despite the complexity and diversity of features at these sites, they still exhibit a high degree of symmetry, axiality, and orthogonality in their layouts. Some relationships appear planned and others appear to be an outgrowth of regular site use. Thus, I suggest that these sites are best explained as representing both the complexities of daily life and the likely spatial practices imposed on the community as a whole by its leaders. It is likely that the idea of such an order was a part of a broader shared idea of the formality of site layouts. While there is much more to explore regarding the observations, this analysis offers some initial thoughts regarding the scale and regularity with which formal elements of site layout were incorporated into the daily lives of the Native Americans occupying the southwestern Gulf Coast.

One issue we have in southwestern Florida is that the terms developed for earthen architecture somewhat impede our ability to accurately describe and compare some of the sites in this region. This is not to say that we cannot use typologies in the region, but we need to be careful about what they obscure and how they might limit our understanding of how these sites formed. Smith's approach offers a complementary way to conceptualize these sites that focuses on multiple aspects of a given site's layout, instead of simply attempting to classify features into categories such as mounds, middens, and plazas. By taking Smith's approach, we are able to examine the degree to which relationships are expressed among features at various sites. Information such as this provides key insights into the types of meaning, such as what architectural arrangements might have meant for a site's inhabitants. Thus, this method has, for this analysis at least, allowed me to bypass a debate that has begun to impede progress and is reaching polemic proportions among some researchers.

As M. Smith (2007:25) states, archaeologists look to "similar buildings, layouts, and other urban features in a series of related cities" so that they can identify a common plan or conception of space. The problem, he goes on to note, is that one needs a large sample to do any real comparison that will result in the identification of similarities. The problem lies largely in the difficulty of mapping such sites in sufficient detail, and such comparisons have been slow in coming into the archaeological literature. This, however, is changing with the ability to accurately map large sites relatively quickly with technologies such as LiDAR. And these technologies are certainly a boon for studies such as this, in which dense subtropical vegetation covering these large sites impedes accurate and rapid mapping using traditional archaeological techniques (e.g., total station mapping). Therefore, I do think that archaeology is at a point of revolution in our ability to deal with questions related to this scale of analysis, whether we are examining large urban sites in Mesoamerica or massive anthropogenic islands along the southern Gulf Coast of Florida.

## Acknowledgments

Many people contributed to my ability to produce this chapter. I thank my good friend and colleague Thomas J. Pluckhahn, whose conversations regarding Smith's ideas were the inspiration for this research. In addition, many individuals helped me refine not only my thoughts regarding the analysis of these sites, but also my abilities to process LiDAR data in GIS. They include John Turck, Thomas Pluckhahn, Ellen Burlingame Turck, Amanda Roberts Thompson, William Marquardt, Karen Walker, Matt Colvin, and Stephen Kowalewski. I also thank Bryan Haley and Duncan McKinnon for their kind invitation to contribute a chapter to this volume. Finally, I would like to thank Mike Smith for his careful reading of the manuscript. His comments helped improve the overall quality.

# 9
# Magnetic Susceptibility for Historical Archaeological Sites and Landscapes

*Daniel P. Lynch and Rory Becker*

Magnetic susceptibility surveys are seldom utilized at historical archaeological sites in North America, but they are increasingly utilized in pre-Contact period archaeological research. In her review of magnetic susceptibility techniques at North American archaeological sites, Dalan (2008) addresses why pre-Contact sites are suitable for these methods. She suggests that earthen features common to pre-Contact sites and the ephemeral nature of some of these have stimulated recent soil magnetic susceptibility studies. Dalan further suggests that geoarchaeology and pre-Contact archaeology are historically allied disciplines; therefore, magnetic susceptibility data are more commonly integrated with other chemical and physical methods routinely used in geoarchaeological studies.

Although the more frequent use of magnetic susceptibility surveys in North America has been on pre-Contact sites, the technique does have some applicability on historical sites. In fact, many of the questions of interest to historical archaeologists may be addressed or enriched by the inclusion of magnetic susceptibility surveys in their research designs and methodology. Since the mid-1970s, historical archaeologists have been encouraged to consider the differential use of space and hubs of activity in research designs geared toward understanding daily life in the past (Fairbanks 1977). This has led to much consideration of site-level land use and activity patterns around homes and structures as a means to negotiate social relations (Beaudry 1986, 1999; Delle 1998; Mrozowski 2010; Mrozowski et al. 1996; Rotman 2003; Rotman and Nassaney 1997).

Mainly because the effects of fire can enhance soil magnetism, magnetic susceptibility surveys have the potential to directly identify activity areas related to domestic or industrial use of space. Historical period sites and landscapes can be materially dense with burned areas and landscape modifications, making magnetic susceptibility surveys a suitable tool for site investigations. Three historical sites from different geographic locations within the United States provide case studies for the use of magnetic susceptibility surveys (see Figure I.1). Fort Bridger, Wyoming, is a multioccupation site historically related to the fur trade, Mormon emigration, and the US Army. The Pointe aux Barques Lighthouse was home to Michigan's first female light keeper and her family. Finally, a historical tar kiln in North Carolina once played a role in the rural industrial production of naval stores.

## Magnetic Enhancement of Sediment

Fine-grained magnetic particles including iron oxides and oxyhydroxides are found in most sediment throughout the world. Magnetite ($Fe_3O_4$) and maghemite ($\gamma$-$Fe_2O_3$) exhibit the strongest magnetism, while others (e.g., hematite) are not very magnetic. Many weakly magnetic iron minerals are very sensitive to changes in their environment and will transform into highly magnetic forms under the right conditions. Soil magnetism can increase or decrease as a result of numerous factors, including heating from natural and cultural fires, anthropogenic sources such as pollution, natural pedological soil processes, and biogenic sources including magnetic bacteria (Dearing et al. 1996; Dearing et al. 2001; Fassbinder et al. 1990; Le Borgne 1960; Linford 2004; Liu et al. 2012; Maher and Taylor 1988; Maki et al. 2006).

Because soil magnetism may vary as a result of human activity, magnetic susceptibility surveys have the potential to speak to the differential use of space within a site. The ubiquitous use of fire by human populations is an important mechanism for the enhancement of soil magnetism. However, evidence suggests that magnetic enhancement might arise from cultural activities other than fire. Most of the topsoil in the world is naturally more magnetic than the underlying subsoil, so the inclusion of topsoil in archaeological deposits might result in localized magnetic enhancement. In addition to the mechanical movement of magnetic soils by human labor, culturally modified soil conditions might encourage inorganic precipitation and/or microbial activity, leading to magnetic enhancement (Linford 2004). Le Borgne (1955) first suggested that magnetic enhancement found in topsoils might arise from the "fermentation" effect, wherein alternating wet and dry cycles would result in the bioreduction of iron in anaerobic conditions and precipitation of magnetite or maghemite. Biogenesis is also a factor, and magnetite-producing bacteria are reported from organic-rich archaeological deposits in higher concentrations than in the surrounding soil matrix (Fassbinder and Stanjek 1993).

## Case Study—Fort Bridger, Wyoming

In 1843, Jim Bridger and his business partner Louis Vasquez established a trading post along the Oregon Trail in southwestern Wyoming (see Figure

I.1). They sold the post to Mormon settlers in 1853 and the Mormons fortified the original log structure with a substantial stone wall. The Mormon settlers continued to operate the trading post through the 1850s, and it catered to Utah-bound immigrants until political conflict arose between the United States and the Mormon settlers. The US Army marched on the location in 1857 and the Mormon settlers burned the trading post, fleeing in advance of the troops. The US Army occupied the remnants of the trading post during that winter and began new construction around the Mormon compound in the spring of 1858.

The US Army used the fort as a base of operations for southwestern Wyoming and northern Utah until it abandoned Fort Bridger in 1890, when the land was transferred to the Department of the Interior. After years of homesteading activity, the state of Wyoming erected the first historic marker at the site in 1914 to commemorate the Oregon Trail. This helped pave the way for Fort Bridger to become a Wyoming State Historic Site when the Wyoming Historic Landmark Commission acquired the property in 1929 (Ellison 1931; Hoagland 2001). The modern Fort Bridger Rendezvous began in 1972 and marked a very new type of activity on the site, in which as many as 300 reenactors may camp on-site and thousands of visitors attend the event over Labor Day weekend each year.

In an effort to assess potential impacts from the activities on the site and to identify subsurface structures related to the military period, the Wyoming State Lands Commission provided funding for a geophysical survey in 2007. The survey was accomplished primarily with a Bartington Grad601 magnetometer (fluxgate gradiometer) and a Bartington MS2-D magnetic susceptibility meter with field coil (Bartington Instruments 2011a, 2011b). Electromagnetic conductivity was also utilized in select areas, although the results are not discussed here (Lynch and Becker 2008). The results of the magnetic susceptibility survey at Fort Bridger show the potential of this method for identifying areas of intensive human activity across a historical landscape (Figure 9.1a). One area of note is near the enlisted men's barracks in the south portion of the survey area. The area around the barracks highlights the contrast in the type of information gained from magnetometry data (see Figure 9.1b) versus wide-mesh magnetic susceptibility data. Underground water pipes associated with three laundries behind the enlisted men's barracks are easily identifiable as linear anomalies in the magnetic gradiometry data, though little direct evidence for the foundations of the barracks can be seen. The magnetic susceptibility data, in contrast, show a high intensity of magnetic soil enhancement behind the barracks. Additionally, a comparison of the magnetic susceptibility data from behind the barracks to the data from the parade grounds shows that there is potentially far more activity represented in the data from behind the barracks. The parade grounds represent formal public and ceremonial space regulated by the discipline of military life. The areas behind the enlisted men's barracks are less public and are out of the line of sight of the command offices and officers' barracks (see Figure 9.1d). The magnetic susceptibility data suggest that the space behind the enlisted men's barracks was used in a much different way than the parade grounds. It is potentially behind the barracks that soldiers would have engaged in all manner of daily activities, including those not approved of by their commanding officers.

Magnetic susceptibility surveys have the capacity to provide direct evidence about human activity areas within a site. The highest concentration of anomalies in magnetic susceptibility data bears a close relation to historical military structures, especially the spaces behind those structures away from more formal spaces. This patterning of anomalies in relation to buildings is mimicked with other structures on the site such as the blacksmith's shop, the fuel storage structure, the bakery, and others not discussed here (see Figure 9.1c, d). Of particular note is the distinct patterning of elevated levels of magnetic susceptibility north of the fuel storage structure versus the relative absence of magnetometer anomalies in the same area from the magnetic gradiometry data. This contrast between the two methods may arise when magnetic soils are thinly deposited yet widely dispersed in the horizontal context as a landscape feature. In our experience, this type of thin yet widely dispersed magnetic landscape feature is best delineated with a magnetic susceptibility survey. The contrast between magnetometry and magnetic susceptibility demonstrates the complementary nature of these two techniques, allowing for different types of interpretations to be made about human activities across large historical sites and landscapes.

## Case Study—Pointe aux Barques Lighthouse, Michigan

The Point aux Barques lighthouse is located in Michigan north of the town of Port Hope (see Figure I.1). The lighthouse was built in 1848 to aid ships in avoiding the shallow reef that extends nearly two miles out from Point aux Barques into Lake Huron. In addition to marking the hazard, the lighthouse served as a land-based navigation point for ships making the turn around the thumb of Michigan into Saginaw Bay. The Point aux Barques lighthouse is listed on the National Register of Historic Places and still serves as a navigational marker for ships passing through the Great Lakes.

Peter Shook was the first light keeper at the Point aux Barques lighthouse. Peter and his wife, Catherine, along with their eight children, lived in the keeper's house close to the light tower. Tragedy struck the Shook family in March 1849 when Peter drowned after his boat capsized en route

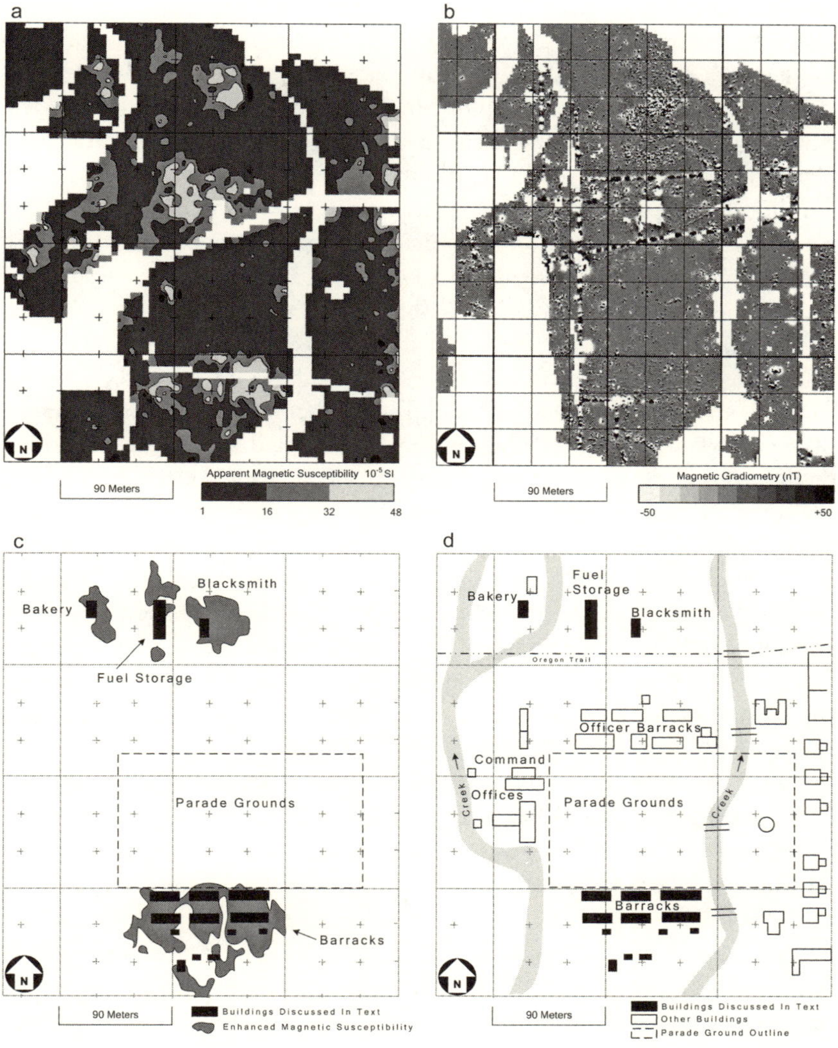


Figure 9.1. Apparent magnetic susceptibility and magnetic gradiometry results at Fort Bridger: (a) apparent magnetic susceptibility results; (b) magnetic gradiometry results; (c) enhanced magnetic susceptibility at the bakery, fuel storage, blacksmith's shop, and enlisted men's barracks; (d) plan of Fort Bridger. Note that the magnetic gradiometry results show historical period subsurface utilities behind the barracks and that the enhanced magnetic susceptibility in the same area indicates high levels of activity in contrast to the low magnetic susceptibility levels in the public/ceremonial space such as the parade grounds.

to Port Huron. Catherine Shook took over the position for her husband, becoming Michigan's first female lighthouse keeper. However, the situation soon went from bad to worse when a fire broke out in the ceiling of the keeper's house and Catherine was badly burned while trying to stop the fire from spreading to the light tower. Although Catherine succeeded in saving the light tower, the family lost nearly everything as the fire completely consumed their home. Catherine and her children resided in the lighthouse until, at the urging of lighthouse inspector Henry Miller, a new home for the family could be constructed. Catherine Shook served as light keeper at Point aux Barques for just two years, until March 19, 1851, when she resigned the position.

The original stone light tower that served as an impromptu home for Catherine Shook and her family was demolished by 1857. A new light tower was built and is still in operation today. There was some debate about whether the new tower and keeper's house were constructed over the foundations of the original tower and keeper's house used by the Shook family. The Pointe aux Barques Lighthouse Society invited archaeologists from Western Michigan University to investigate this question in 2003 (Nassaney 2003).

The 2003 geophysical survey at Pointe aux Barques utilized twin-probe electrical resistance (TR Systems, Ltd. 2002) and MS2-D field coil magnetic susceptibility (Bartington Instruments 2011b). Electrical resistance data were collected with a 50 cm mobile probe array on a 0.5 x 0.5 m grid (four readings per $m^2$). Topsoil magnetic susceptibility data were collected on a 1 x 1 m grid (one reading per $m^2$). The electrical resistance results show the foundations of the original stone tower as a doughnut-shaped anomaly of high resistance near the center of the image (Figure 9.2a, c). Additionally, the original 1847 keeper's house is identified in the results of the resistance survey. Systematic excavations verified these anomalies as the original light keeper's house and light tower (Nassaney 2003).

While providing evidence for the location of the 1847 structures was of primary concern during the 2003 investigation, the geophysical survey results provide a wealth of additional information that can be used to aid site interpretation. Although the results of the electrical resistance survey locate the lighthouse tower and keeper's dwelling precisely, the data provide little evidence of the domestic life and tragic history surrounding the Shook family's tenure. Besides identifying the location of the two 1847 foundations, the resistance data provide information mostly about modern utilities and walkways maintained by the park (see Figure 9.2a, c).

The magnetic susceptibility data from the Pointe aux Barques lighthouse survey show potential activity areas related to the professional and domestic life and eventual destruction by fire of the original 1847 keeper's

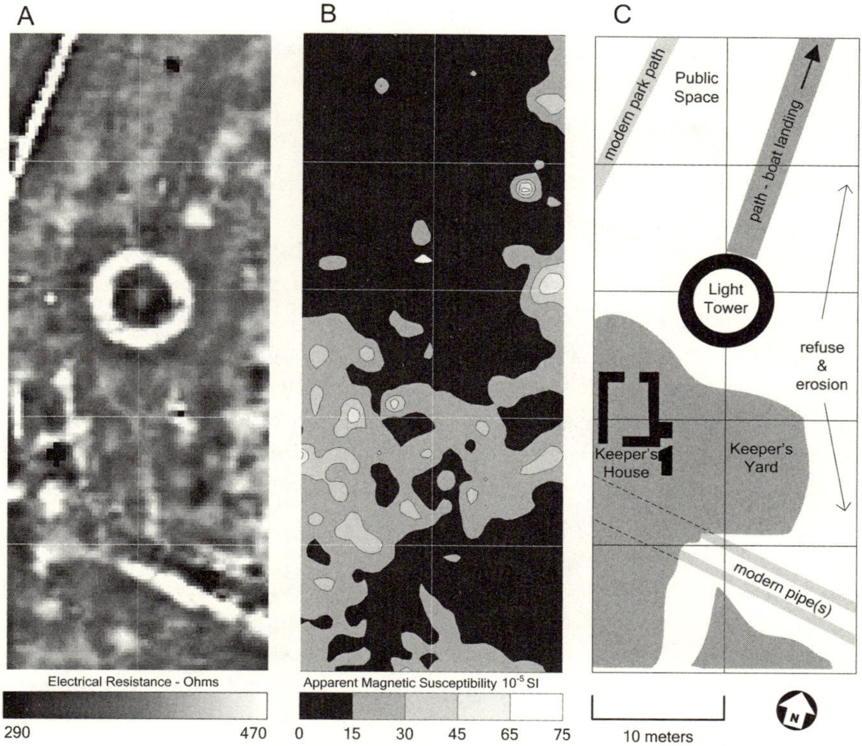


Figure 9.2. Electrical resistance and apparent magnetic susceptibility results for the Pointe aux Barques Lighthouse: (a) electrical resistance results show the 1847 foundations for the lighthouse tower and the keeper's house; (b) magnetic susceptibility results show higher levels of magnetic susceptibility in the area of the 1847 keeper's house (this contrasts with lower levels of magnetic susceptibility in the nondomestic space around and north of the light tower); (c) interpretation of magnetic susceptibility and electrical resistance results.

house (see Figure 9.2b, c). The area around the original light keeper's house in the southwestern part of the survey exhibits a higher apparent magnetic susceptibility than surrounding areas. This pattern is what one would expect to find over a domestic residence that was lost in a catastrophic fire. In contrast, the apparent low magnetic susceptibility found north of the 1847 light tower foundation suggests a level of professional maintenance, or minimally, that different activities occurred here. The area north of the light tower was public space incorporating a boat landing for visitors, including managing government officials who would likely have arrived at this location by boat. The magnetically low pattern north of the light tower contrasts sharply with the apparent high magnetic susceptibility of the keeper's yard to the south of the light tower. Finally, other areas of high magnetic susceptibility are located along the very eastern edge of the survey at a cut bank that drops off sharply to the edge of Lake Huron. The eastern edge of the survey might contain refuse, as some of these data appear to originate from small metallic sources.

The magnetic susceptibility results suggest that from 1848 until 1857, early occupants of the site, including the Shook family and later light keepers, conducted much of their private household business in the area south of or behind the lighthouse tower (from the public perspective). The pattern of lower magnetic susceptibility to the north of the light tower suggests professional management of this public space, potentially maintained and delineated from the private spaces utilized by the light keepers and their families. When combined with the precise locational evidence provided by the electrical resistance survey, the magnetic susceptibility results are suggestive of the spatial patterning of human behavior for occupants of the original Pointe aux Barques light keeper's house.

## Case Study—Tar Kiln, Clear Run, North Carolina

The production of tar, pitch, charcoal, rosin, and spirits of turpentine is a historically important part of rural industry in the southeastern United States. These five products derive primarily from processing the wood and gum of the longleaf pine (*Pinus palustris*) and other southern pine trees. Tar and charcoal are produced in earth-covered kilns, while pitch is a later refinement of the raw tar product. Rosin and spirits of turpentine are products of the distillation of pine gum recovered from live trees, which is an entirely different process unrelated to tar extraction from a kiln. Tar and pitch were vital to the naval stores trade. They were the essential waterproofing agents used in the wooden shipbuilding industry. The immense coastal pine forests of the Carolinas and Georgia became the primary center of this industry. By 1760, the lower Cape Fear River region in North Carolina was the largest supplier of naval stores in the British Empire (K. Robinson 1997; J. Williams 1935).

The construction and operation of a tar kiln have multiple steps ranging from procuring fuel to constructing the kiln, covering the kiln with earth, igniting and tending the fire, and finally collecting the tar and charcoal (M. Harmon and Snedeker 1997:147–148). Of importance to the consideration of a geophysical survey, tar kilns were covered with earth to reduce the oxygen atmosphere and control the intensity of the fire.

Based on foundational research in soil magnetism, oxygen-reducing heating conditions, such as those employed in tar kilns, are known to enhance soil magnetism (Tite and Mullins 1971). In other words, heat as well as reducing and oxidizing conditions is needed for magnetic enhancement (Tite and Mullins 1971). Once the tar is boiled out of the wood under low oxygen conditions, charcoal is a secondary product available to recover (M. Harmon and Snedeker 1997:151; Joseph et al. 2004:119). When the kiln is reopened to recover the charcoal, oxygen is reintroduced into the kiln, providing reoxidizing conditions for potentially even greater magnetic enhancement of the soils.

The primary research question driving this geophysical survey was whether a geophysical survey could noninvasively classify and identify tar kilns by their type, including size, shape, configuration, technology, and other indicators of site integrity. Site integrity and the overall level of preservation at tar kiln sites are important considerations when assessing basic heritage management strategies or eligibility for the National Register of Historic Places (Joseph and Botwick 2009:165–172). A recent government-funded heritage management summary of rural industries in Georgia and the Carolinas suggests that "ideal sites for research into tar making would be kilns with clearly defined shapes or styles and examples that contain chronological information. Information on the basic size, shape, and location of kilns would also be important" (Joseph and Botwick 2009:177).

The tar kiln featured in this case study is located on private property at Clear Run (Garland), North Carolina, in the lower Cape Fear River region (see Figure I.1). The kiln is an earthwork feature shaped like a doughnut. It is circular, with a distinct depression in the center and a ring trench along the outside perimeter. The property is National Register eligible and is considered "the most complete and important of the small trading centers in the county" (National Register of Historic Places 1986). Other archaeological resources on the property associated with the naval stores industry include the remains of the steamboat *A. J. Johnson*, a blacksmith's shop and cooper's shop (Brady et al. 1998), a turpentine distillery, a warehouse, and at least three tar kilns.

The fieldwork at the Clear Run Kiln includes two systematic topsoil magnetism surveys, a magnetometer survey, a topographic survey, and a down-hole magnetic viscosity survey (Figures 9.3 and 9.4). In addition, twin-probe electrical resistance and electromagnetic conductivity surveys were completed but are beyond the scope of this chapter. Magnetic susceptibility was measured with the MS2-D coil (Bartington Instruments 2011b) and magnetic viscosity with the MVM1-D field coil (Pulsepower Developments 2009). Twenty small-diameter bore holes (2.54 cm), separated at 1 m intervals along a single linear line (A′–A′), were investigated at 2 cm sample intervals with the MVM1-H down-hole magnetic viscosity probe.

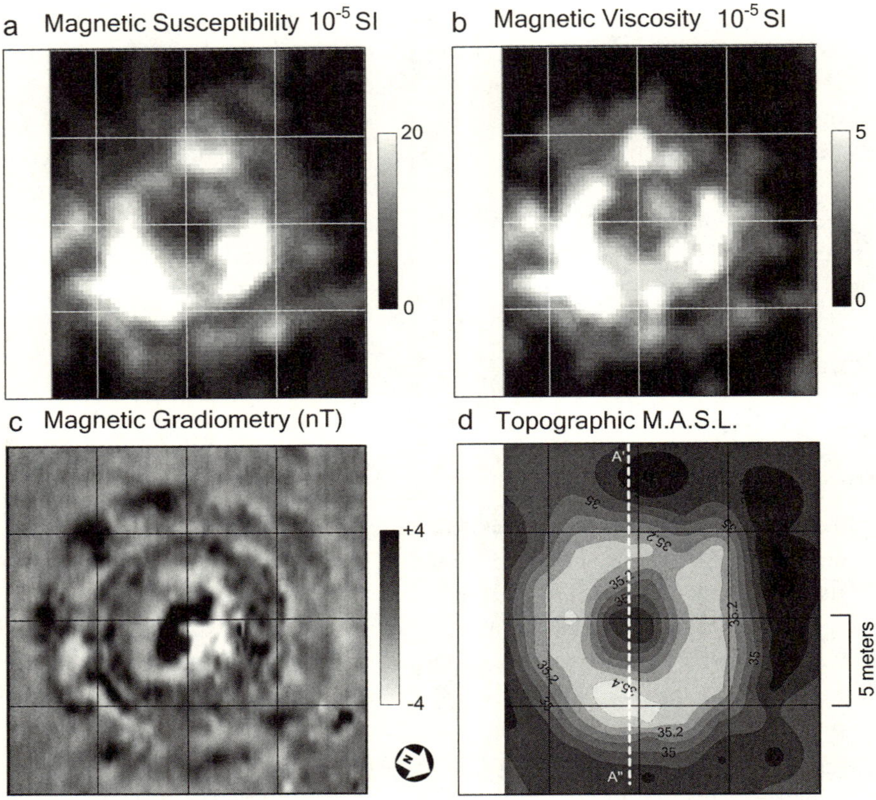


Figure 9.3. Multi-instrument geophysical survey results from the Clear Run Tar Kiln site: (a) magnetic susceptibility; (b) magnetic viscosity; (c) magnetic gradiometry (potential evidence for oxygen regulation seen as alternating high and low anomalies along the outer edge); (d) topographic contour map of the kiln's surface. The line from A′ to A″ represents the location of the bore-hole line for the down-hole magnetic viscosity survey (see Figure 9.4).

The magnetometer data were collected with the Bartington Grad601 (Bartington Instruments 2011a), and the topographic survey data were collected with a rotating laser level.

The topsoil magnetic susceptibility and magnetic viscosity surveys were conducted quickly with low data density (one reading per $m^2$) yet provide details unavailable to the naked eye. The upper right-hand corner of both anomalies appears as the open end of a *C*, potentially indicating where the tar collection pipe flowed out of the kiln (see Figure 9.3a, b). Interestingly, both the topographic map and the magnetometer show a similar *C* shape opening toward the upper right, indicating that this anomaly has some merit.

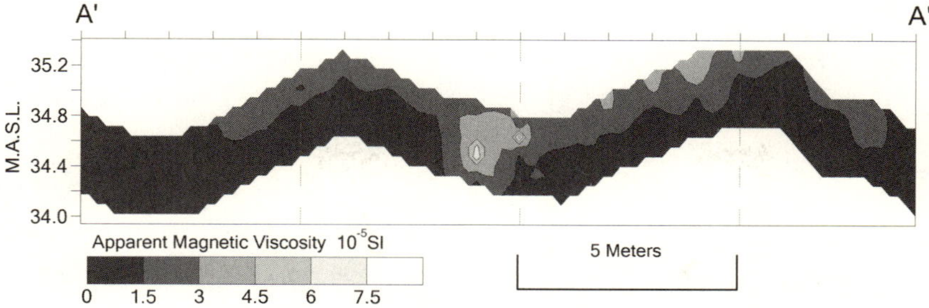


Figure 9.4. Topographically corrected down-hole magnetic viscosity profile (see Figure 9.3d for location). Among other things, the results show the shape and depth of a central buried anomaly with high magnetic viscosity.

The magnetometer survey was conducted at a much higher resolution (16 readings per $m^2$) than the other methods. At the center of the kiln is a strong response, potentially of thermoremanent origin. These types of anomalies are expected and are good indicators that the feature has not been significantly disturbed.

An unexpected but welcome result from the magnetometer survey is found along the outside edge of the kiln anomaly. A series of alternating high and low (dark and light) anomalies repeat and give a buzz saw appearance to the outer edge of the kiln. This alternating chain of anomalies is found where the ring trench meets the outside earthen kiln wall (compare Figure 9.3c, d). One explanation might be that this alternating series of high and low magnetism is related to controlling oxygen flow into the kiln. At places where oxygen would be allowed to flow in, we would expect more heat and higher-intensity magnetic anomalies. The pattern visible in the magnetic data might represent "undulating floors with alternating high and low spots. Undulating trench floors may have resulted from the practice of opening and closing holes in the kiln sides to promote even burning and prevent complete oxidation" (M. Harmon and Snedeker 1997:151).

The location of the down-hole magnetic viscosity survey in relation to the tar kiln is noted on the topographic map with starting point (A′) and ending point (A″) (see Figure 9.3d). The down-hole data clearly delineate what appears to be an internal area of significant high magnetism, slightly off center from the kiln (see Figure 9.4). This large magnetic viscosity anomaly likely results from an internal feature that extends down about 65 cm below the ground surface.

The evidence from the magnetic viscosity down-hole survey, along with the other geophysical evidence, suggests that the Clear Run tar kiln remains intact with a high degree of structural integrity. While metal detector surveys and LiDAR are already being recommended to find short-term labor camps and associations between these site types, the results reported here suggest that geophysical surveys may also assist in refining the typologies currently used to classify tar kiln sites (Espenshade et al. 2012; Southerling 2013).

## Conclusion

Magnetic susceptibility surveys have the potential to highlight areas of human activity at historical archaeological sites and landscapes. The three case studies presented here demonstrate a high potential to inform the types of research questions commonly asked in historical archaeology. The magnetic susceptibility results at Fort Bridger provide insight into differential use of space across a historical landscape. The space behind the enlisted men's barracks suggests a high amount of activity in contrast to public areas in plain view of the officers' quarters, which did not share this pattern (e.g., parade grounds). This information was collected inexpensively and also very rapidly; however, it opens new avenues to form research questions about the types of activities taking place behind the enlisted men's barracks as compared to the types of activities occurring in full view of the officers' quarters.

The 2003 geophysical survey at the Pointe aux Barques Lighthouse site identified the location of the 1847 lighthouse tower and keeper's house, which was the primary goal of the site investigators. However, results of the magnetic susceptibility survey offer a new direction for research questions about the spatial delineation maintained between the public and private lives of the early occupants of Pointe aux Barques Lighthouse.

At Clear Run, North Carolina, magnetic susceptibility and magnetometry surveys of a tar kiln provide interpretation of the kiln's technological aspects including size, shape, typology, oxygen regulation, and tar collection. Down-hole magnetic viscosity testing provides clear evidence for intact subsurface features within the tar kiln. These results suggest that rapid typological classification is possible with geophysical surveys at tar kiln sites, which would aid heritage management efforts on similar sites related to the industrial history of the area.

# 10
# Anthropologically Focused Geophysical Surveys and Public Archaeology

## Engaging Present-Day Agents in Placemaking

*Edward R. Henry, Philip B. Mink II, and W. Stephen McBride*

On January 1, 1863, notorious Civil War raider and Confederate general John Hunt Morgan charged through Taylor County, Kentucky, on his way south to Tennessee during the Christmas Raids of 1862 to 1863. During his travels through the county he visited a small stockade recently constructed by members of the Union Army, situated at the bluff crest of a ridge overlooking the Green River. This stockade was built to protect a vital bridge over the river where it contorts to form what is known as Tebbs Bend (Figure 10.1; see also Figure I.1). The Green River covered bridge was an important link in a supply line that provided food and munitions to Union troops across Kentucky. After encountering the site, Morgan ordered the bridge and stockade burned before continuing on his holiday raiding spree (Gorin 2006:88).

General Morgan would return from Tennessee in late June or early July 1863. During his time away, Colonel Orlando H. Moore, the newly arrived Union officer overseeing the Tebbs Bend area, had begun repairs of the stockade and the Green River Bridge with infantrymen from Michigan, New York, and Kentucky (Gorin 2006:152–154). General Morgan returned as part of his "Great Raid"—a series of incursions into Indiana and southeastern and eastern Ohio near the border with West Virginia and Pennsylvania. Because Colonel Moore was aware of General Morgan's advances back into Taylor County, he was able to plan a defensive strategy that utilized the unique topography formed by the Green River at Tebbs Bend. Particularly advantageous was a tight restriction, or "choke point," on the ridgetop that sat at a lower elevation than the rest of the landform, located almost a mile from the stockade. Colonel Moore ordered a rifle pit excavated at this point (see Figure 10.1). Approximately 100 yards upslope from this entrenchment a fallback locale and main defensive position were es-

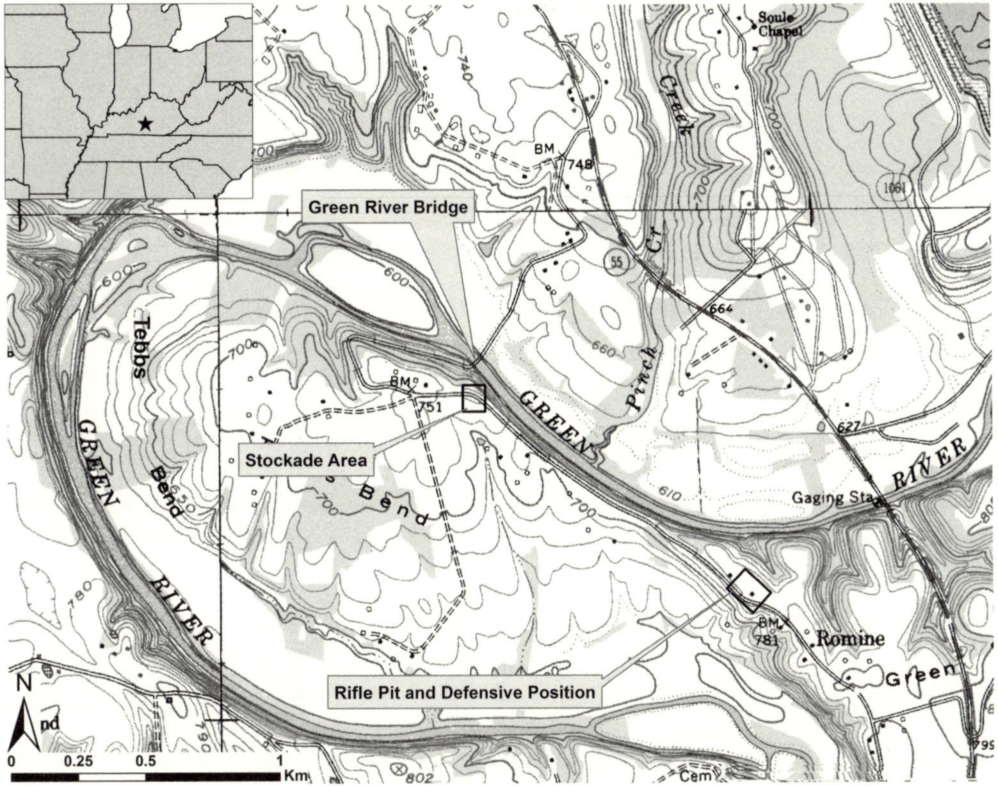


Figure 10.1. Location of the Tebbs Bend Battlefield on the Cane Valley 7.5 minute USGS topographic quadrangle, Taylor County, Kentucky. Survey areas highlighted. Cem, cemetery; BM, benchmark location from USGS. (Imaging data from Kentucky Division of Geographic Information and USGS.)

tablished that consisted of at least an abatis and possibly a log breastwork (Gorin 2006:168–173).

General Morgan and his men opened fire on the Union Army at around 6:30 on the morning of July 4, 1863. With difficulty, they were able to capture the rifle pit. This advancement required the forward Union troops to retreat to their timbered primary defensive position. This rendered the rifle pit and artillery useless to the Confederates because of their topographic setting well below the main defensive position. From this position, Union soldiers under the direction of Colonel Moore held off General Morgan's advance because of the inability of the Confederates to flank the Union position in this topographic setting (Gorin 2006:187–194, 198). By 10:30 a.m. General Morgan gave up and moved his soldiers in another direction to avoid the Green River Bridge and Colonel Moore's soldiers. This unique use of space provided the advantage needed for approximately 250 of Colonel Moore's lightly armed Union troops with no artillery to hold off almost 2,500 Confederate soldiers and four pieces of artillery under the command of General Morgan. The Battle of Tebbs Bend saw the Union maintain control of the Green River Bridge and the newly reconstructed stockade, both of which endured throughout the rest of the war. This feat of military tactics remains one of the most impressive narratives from Civil War battles fought in Kentucky.

In its contemporary setting, rural farmers and householders of Taylor County occupy the Tebbs Bend ridgetop. Some have grown up playing on portions of the battlefield. However, over time, and certainly by the 1950s, the placement of Union defensive features and the landscape of military strategy linked to the battle of Tebbs Bend were no longer visible. As a result, the battle was becoming lost to the local imagination. The slow but constant advance of modern agricultural practices and rural road improvements were continually diluting the memory of Colonel Moore's unlikely triumphs against the Confederacy. As we will see, some of Tebbs Bend's present-day occupants were losing touch with the landscape and where these events transpired. Moreover, the weakening memories related to the battle were leading to the initiation of what Relph (1976) refers to as "placelessness," or loss of place identity. Placelessness can be reversed by providing local communities a mnemonic by which they can attach themselves to local landscapes and engage in the practice of placemaking. We argue here that archaeological geophysics can become such a mnemonic if results are utilized in the correct ways.

In this chapter we make a case for the use of geophysics as a tool to engage public and local communities. We believe that geophysics can help descendant communities, and those Wright (2015) refers to as "resident communities," better understand the past and restore their connection to places, or create connections where none previously existed. We present a case study showing how resident communities can engage with archaeologists employing geophysical methods; however, archaeologists working on prehistoric sites (see contributions to this volume) could benefit from using their geophysical data sets in ways that provide descendant communities with a means to participate in present-day placemaking through a better understanding of the landscapes that their ancestors inhabited.

Archaeological geophysics provides a tangible representation of features on and under the landscape, as well as the primary data necessary to explore spatial relationships between these components. Both have been championed as a way to investigate human activity on the landscape when considered from perspectives in anthropological archaeology (Kvamme 2003a; V. Thompson et al. 2011). Thus, it provides human agents a mechanism through which they can negotiate the significance of their surrounding environment, initiating present-day placemaking. We open this chapter with perspectives on geophysics, public archaeology, and placemaking and collective memory. We then summarize the Tebbs Bend project, its goals, and results. Next, we offer insights into how our work has affected perceptions of place through summaries of three interviews we conducted with residents of the Tebbs Bend area after the conclusion of our project. The responses to our interviews show that local agents became reinvigorated with the past and reconnected with their sense of place on the battlefield through our work. We close the chapter with a call for archaeological geophysicists to organize their research around a focus that might produce similar outcomes for both resident and descendant communities.

## Geophysics, the Public, and Place on the Landscape

Geophysics has come a long way since becoming a common prospection tool for North American archaeologists (J. Johnson 2006). Kvamme (2003a) provided the seminal call for landscape-scale anthropological analyses using these methods. Since then, scholars have been pushing more geophysical practitioners to explore anthropologically focused questions within archaeology using their data (Conyers 2010; Conyers and Leckebusch 2010). Recently, V. Thompson et al. (2011) laid out specific ways these methods can be adopted within an anthropological research framework to answer questions related to variation in architectural construction, continuity and discontinuity on the landscape, natural versus cultural modifications, and regional-level patterning. Practitioners have answered these calls, as evidenced in a 2014 special edition of *Archaeological Prospection* (vol. 21). However, one major benefit of archaeological geophysics has been overlooked: it is a tremendous benefit in the context of public archaeology and a great tool with which to engage the public.

Public archaeology affords us the opportunity to develop relationships with, and effect change in, contemporary communities that have direct and indirect connections to the past (Little and Shackel 2007; Okamura and Matsuda 2011; Stottman 2010). It provides individuals and communities with the knowledge needed to become honorable stewards of our heritage. Practicing archaeology in a public sphere advances the notion that we must understand, conserve, and preserve the past for the future, promoting civic responsibility along the way. Here, geophysics has a role to play as we move forward using these technologies in the public realm. Like the anthropological focus advocated in research-based projects, public archaeology can also benefit from geophysical surveys. Public-oriented research agendas can be improved by contextualizing spatial relationships with the built and natural environment (*sensu* V. Thompson et al. 2011), but geophysics also offers a technologically appealing channel to the public. With geophysical data sets in hand and anthropological interpretations grounding our reasoning, we are poised to put forward and discuss narratives regarding how and why past humans engaged in landscape alterations. When geophysics imparts tangibility to archaeological sites and our communication with the public, it materializes what is beneath our feet, unseen. This process provides a mechanism through which the public can engage their imaginations and negotiate personal relationships to places, events, and people in the past as they relate to the present. We draw on the popular concept of placemaking to describe this process.

Placemaking might be defined broadly as the process whereby places are developed and managed physically with respect to how the spaces they define are imagined and interpreted (Basso 1996; Beatley and Manning 1997; Berry 2011; Tuan 1977). Mechanisms of placemaking not only relate to the development or remodeling of particular physical places but also include historical events, experience, imagination, and movement through space (Basso 1996:107; Nabokov 2002:145; Tuan 1977:7, 146). The residents of Tebbs Bend engaged in placemaking through a palimpsest of Civil War history, local knowledge, imagination, and individual experience.

Particularly relevant to our case study are the writings of Wendell Berry. Referring to American perceptions of the Civil War, he writes, "If what we see and experience, if our country, does not become real in imagination, then it never can become real to us, and we are forever divided from it. . . . Imagination is a particularizing and local force, native to the ground under foot" (Berry 2011:32). Berry's perspectives on placemaking and its connections to history, individual experience, and imagination coincide with many threads in memory studies. For instance, Halbwachs's (1992) notion of "collective memory" describes how the currents of continuous thought function to retain the past only as long as the consciousness of a group is capable of keeping the memory alive—a process always informed by, and negotiated in, the present (Halbwachs 1992:142–143). This reinforces Berry's consideration of placemaking as a process bounded by degrees of social localization and may be one reason why modern understandings of the Tebbs Bend landscape became somewhat diluted through time. Halbwachs (1992:52) writes that collective memories are only part of a common group we sustain relationships with on a regular basis. For the Tebbs Bend community, knowledge of where particular events of the 1863 battle took place was only as accurate as what was told to them by parents and other relatives, who were informed by those who lived on, or knew people who lived on, the landscape near the time of the battle.

This may be one reason why our geophysical surveys were successful in helping residents reminisce about Union defensive features and their spatial relationships. Through this process, new experiences and ties to the past were introduced into the collective memory of the community. These entangled currents of thought served as a basis of contemporary placemaking at Tebbs Bend. The results of our surveys provided tangible evidence of where particular aspects of the Civil War battle took place. In turn, the reconstructed memories of these particular places informed the broader remembering of the battle and where particular events corresponded with individual properties and their owners.

## Resident Communities and the Tebbs Bend Project

Archaeological research at the Tebbs Bend Battlefield coincided with ongoing preservation efforts of the Tebbs Bend–Green River Bridge Battlefield Association (TBGBA). Preserving an archaeological site of this scale requires a thorough understanding of where specific events related to the battle unfolded. Such knowledge is expected because informative signage is needed to guide tourists through the battlefield, linking their experience in the present with what happened in the past. This necessitated a detailed review of archival records, Civil War history, and informant interviews, combined with geophysical surveys and limited excavation. What we discovered through geophysical surveys gave substance to Union Army accounts of the tactics and events mentioned at the start of this chapter. Moreover, it provided local landowners with a better understanding of specific events that transpired during the battle.

Our surveys focused on delineating three particular defensive positions related to the battle: the Union's Green River stockade, the Union rifle pit situated at the "choke point" on the ridgetop, and the Union's primary defensive position where Colonel Moore's soldiers fell back after abandoning the rifle pit. An oil painting of the Green River Bridge and stockade by Private Edward Henry of the Seventy-Ninth New York Infantry (Figure 10.2a) and a sketch map of the ridgetop by Lieutenant Michael A. Hogan of the Eighth Michigan Infantry in 1863 (Figure 10.2b) were the two most influential primary documents that guided the placement of our surveys.

Lieutenant Hogan's 1863 sketch map of the ridge depicts the rifle pit as a zigzag line with the words "Earth Work" above it. It is situated beside a sharp bend in the road that runs through the battlefield. This area falls in the front yard of landowners Raymond and Anita Brown. To the rear of this feature are marks with the words "Fallen Timber" written over them. Gorin's (2006:168–173) work with primary source data related to the battle places this timbered defensive position approximately 91 m (100 yd) behind the rifle pit on property owned by Betty Williams. Private Henry's oil painting of the Green River Bridge and stockade shows the stockade situated on the bluff promontory directly above the bridge. Lieutenant Hogan's sketch map depicts the stockade in a similar locale. This area falls within a small cattle pen owned by Robert and Suzanne Phillips.

Based on these archival sources we organized multi-instrument geophysical surveys in the Browns' front yard, where the rifle pit was purported to be, and a gradiometer survey in a large field to the west owned by Betty Williams. In addition, multi-instrument geophysical surveys were conducted in the Phillipses' cattle pen and in an open field south of Tebbs Bend Road. We collected magnetic susceptibility, electrical resistance, and magnetic gradiometer data over the rifle pit. Since we expected to map a metallic debris field associated with the stockade, only magnetic gradiometer and electrical resistance data were collected over that area before excavation. The resistance data we collected in both locales showed no clear anomalies related to any Civil War features. As a result, we omit discussion of those data sets below.

Ground-penetrating radar (GPR) data were collected over the stockade after excavation. The reason for the gap in time between our initial surveys and the GPR survey was related to our initial expectations for results. Even though we had historical evidence to suggest the stockade was built in this general locale, the magnetic gradiometer and resistance data collected over this area were quite noisy and showed no debris field, only a faint magnetically enhanced ring-shaped anomaly. This led us to believe that agricultural practices associated with the cattle pen had significantly disturbed the sub-

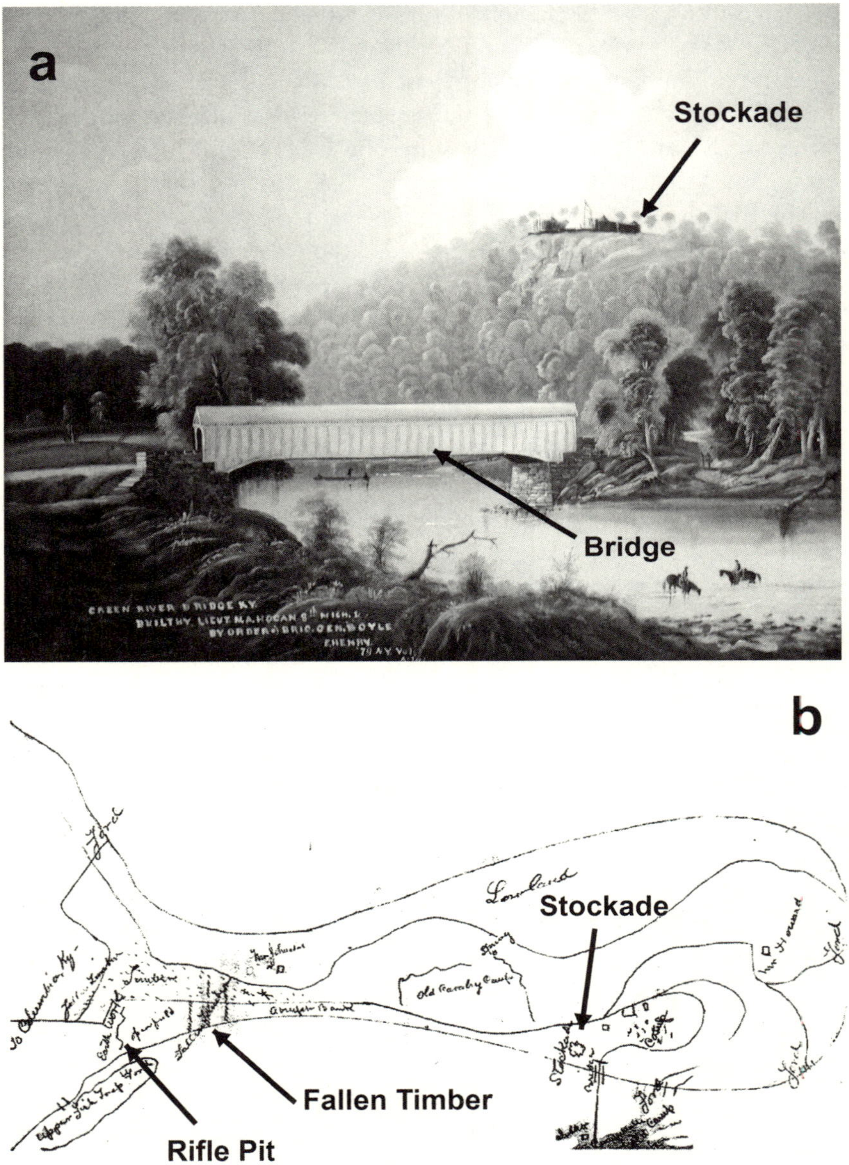


Figure 10.2. (a) Oil painting of the Green River Bridge and Union stockade by Private Edward Henry, Seventy-Ninth New York Infantry; (b) sketch map of Union defensive features on Tebbs Bend ridge by Lieutenant Michael A. Hogan of the Eighth Michigan Infantry (private collection).

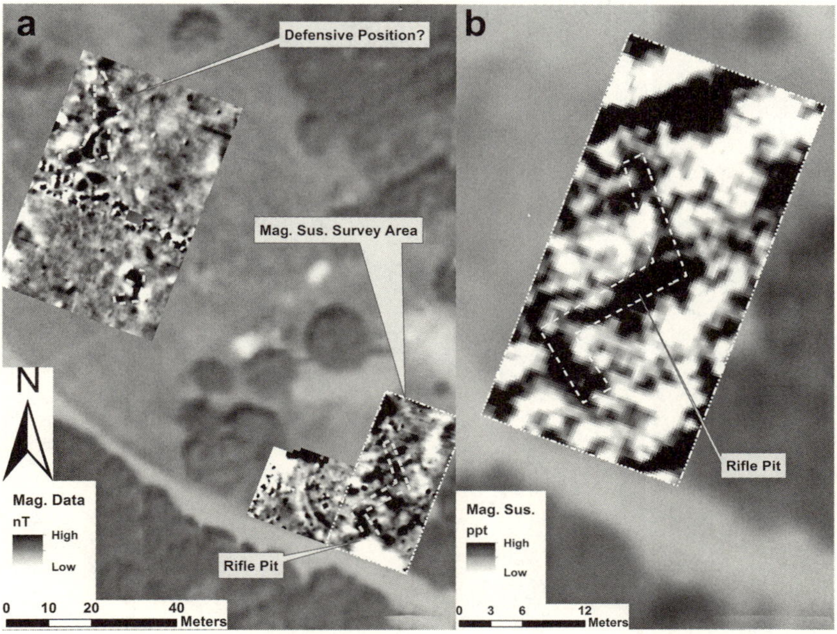


Figure 10.3. (a) Magnetic gradiometry data from the rifle pit area. Data were collected with a Geoscan FM256 and processed in Geoplot 3.0 using zero-mean traverse, destagger, and interpolation functions. (b) Magnetic susceptibility data from the rifle pit. Data were collected using a Geonics, Ltd. EM38A and processed in Geoplot 3.0 using zero-mean traverse and interpolation functions. (Imaging data from the US Department of Agriculture, National Agriculture Imagery Program, Farm Service Agency, Natural Resources Conservation Service, Digital Globe, Inc.)

surface here. After limited excavations revealed intact portions of the stockade, we allocated enough time (a few months) for excavated sediments to settle and then conducted the GPR survey to map unexcavated portions of the stockade.

The results of our surveys near the Brown property showed a zigzag anomaly comprising high magnetic values in both the gradiometer and magnetic susceptibility data (Figure 10.3). West of this area was another high magnetic signature that spatially corresponded to the Union infantry's defensive location built from "felled-trees." Limited excavation of the rifle pit anomaly included two mechanical trenches oriented to bisect the feature. Profiles of the trenches revealed that the remnants of the rifle pit were very shallow (ca. 20–25 cmbs) and were refilled with unconsolidated and intermixed fill of a dark brownish-red clay loam. This displacement of ferrous-

Figure 10.4. Magnetic gradiometry data from the stockade area. Data were collected with a Geoscan FM256 and processed in Geoplot 3.0 using zero-mean traverse, destagger, and interpolation functions. (Imaging data from ESRI, Digital Globe, Inc., GeoEye, Earthstar Graphics, US Geological Survey, AEX, Imagery, Getmapping, Aerogrid.)

rich unconsolidated clay fill likely represents the matrix responsible for the elevated levels of soil magnetism.

Mechanical trenching excavations at the possible fallback location of felled trees identified no buried structural remnants. Nevertheless, we argue that this is still a good candidate for the main defensive position. If this protective locale comprised only numerous horizontally stacked trees, there may be an induced magnetic signal from its construction, as well as a general anthropogenic enhancement of the topsoil from the disturbance that is sensitive to a gradiometer but does not impact the subsurface enough to visibly contrast with the local B and C soil horizons. Thus, it may be detectable geophysically but invisible to the human eye.

Survey results from the Phillipses' farm and cattle pen show a faint semicircular magnetically enhanced area (Figure 10.4). Exposure of this feature via mechanical stripping revealed that it was a wall trench. Artifacts temporally and functionally related to military roles, and dating to the Civil War, were recovered from an excavated sample of the trench. The artifacts,

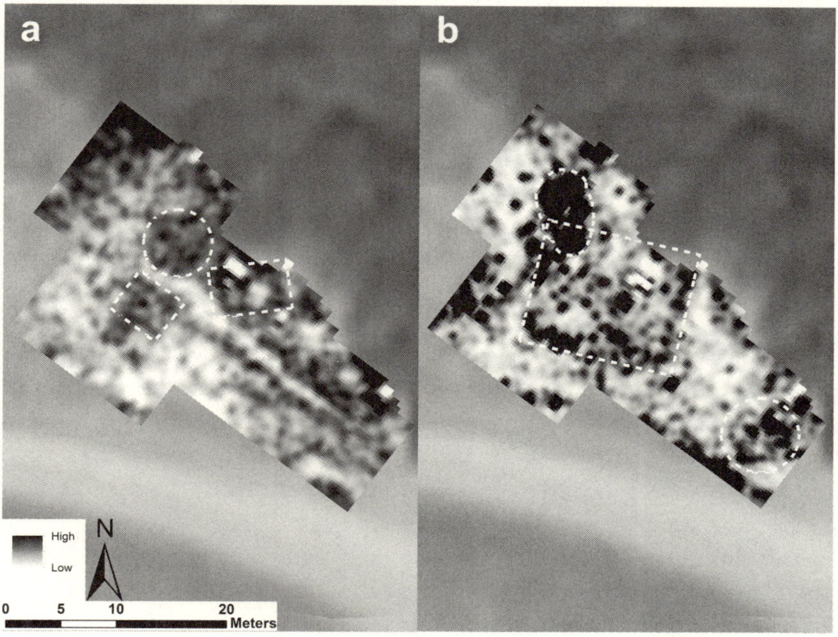


Figure 10.5. (a) GPR data from the stockade area at a depth of approximately 7–16 cm below the surface. All GPR data were collected in the *X* and *Y* directions using a Mala Geoscience Ramac GPR CU II Geo System with a 500 MHz shielded antenna along transects spaced 0.5 m apart. Processing was conducted in GPR-Slice using the default bandpass filter and a 3 x 3 low-pass filter. (b) GPR data from the stockade area at a depth of approximately 40–49 cm below the surface. (Imaging data from the US Department of Agriculture, National Agriculture Imagery Program, Farm Service Agency, Natural Resources Conservation Service, Digital Globe, Inc.)

in addition to the spatial configuration of the trench, confirmed its association with the stockade and suggest it was one of the bastions. A few months after excavations, the GPR survey was conducted over this area in both the *X* and *Y* directions to delineate, in high resolution, features related to the organization of the Green River stockade. Processing the GPR data required numerous filters because of the bedrock deflated soil profile on the Tebbs Bend bluff top. Nevertheless, we were able to define rectilinear high-amplitude reflections approximately 16 cm below the surface, the depth at which the circular bastion was encountered in excavation (Figure 10.5a). Because of the spatial arrangement and depth of these reflections, we argue that they are likely associated with the ground plan of the second stockade at Tebbs Bend. We detected an additional faint rectilinear pattern of reflections approximately 33 to 62 cm below the surface that was offset from the angle of the second stockade construction (see Figure 10.5b). We believe this pattern may relate to the placement of the original Green River stockade burned by Morgan's men (Gorin 2006:88).

Our geophysical surveys were highly successful in identifying two of the three Union defensive positions for the Battle of Tebbs Bend. It is possible that the third "felled tree" position was identified geophysically. While we find our results very encouraging in the context of Civil War archaeology at the site, during the process of collecting and processing our geophysical data we began to notice processes of present-day placemaking through our interactions with the landowners with whom we worked. At both battlefield areas, local landowners began coming out to greet us, talk with us about what we were doing, and ask how our machines operated. Most geophysical archaeologists likely have similar experiences to recount. However, in this case locals began engaging us with past stories and myths from their ancestors about growing up on or near the battlefield.

In some instances, our data confirmed stories they had been told as young people; in others we contradicted what landowners thought they knew of their immediate cultural surroundings. Through these situations we realized that the process of present-day placemaking was materializing through an engagement with geophysical data. Ties to the past were being made, changed, and reinforced through the intersection of historical and geophysical perceptions of the landscape dating from 1863 to the present. This was an excellent opportunity to explore more fully the role archaeological geophysics had in perceptions of place and the process of placemaking. In what follows we recount phone interviews recently conducted with three individuals who were deeply tied to the project, either through owning a portion of the battlefield we surveyed or by having a direct tie to it through the TBGBA. We hope to show how these agents incorporated our geophysical results into their ongoing negotiation of place at Tebbs Bend.

## Local Narratives and the Revitalization of Place

As mentioned, our work was carried out over two primary areas of the Tebbs Bend Battlefield. These surveys were carried out on three private properties. However, we interviewed the owners of two properties and the director of the TBGBA to understand whether our work had impacts on people not immediately living on the Tebbs Bend landscape. We found that even for those not owning land at Tebbs Bend, a deep interest in the history of Taylor County provided the context necessary to be influenced by geophysical results when it came to perceptions of place. In an attempt to structure our interviews along a similar format we began by mentioning who we were and asking interviewees how they were doing. We then moved to asking about Tebbs Bend generally, not just the battlefield site. After we felt comfortable with the flow of conversation we began asking directed questions regarding our survey work and the results. This methodology was followed for all three interviews in an attempt to make this ethnographic component of our project systematic.

Betty Jane Gorin-Smith is an independent historian and author on the Civil War in Kentucky, who also taught history at a local high school and a university in Taylor County. She serves as the director of the TBGBA. When we interviewed Betty Jane we began by asking her how things were generally going with her and her husband. We then discussed our invitation to write this chapter, which sparked conversation regarding the results of our work. In fact, she asked us when we were going to get more funding and do more work. Our conversation provided the opportunity to ask her simple questions related to our project, such as "What did you think of the technology and the instruments we used?" or "What did the results of our geophysical survey make you think about Tebbs Bend?" These two questions became systematic throughout our interviews.

Betty Jane began with normal responses about how our work raised more curiosity about the site and how it verified the historical record. However, she then began telling us that our work had instilled a particular notion of sacredness to the area surrounding the rifle pit and the main Union defensive position. While a highway marker has been positioned in the road bend where the rifle pit was thought to be, our work was the first to confirm its actual location. In turn, this situated the open field behind it as the place where approximately 100 people lost their lives (Gorin 2006:208–213). Betty Jane mentioned that every time she drives this section of the road through the battlefield she is reminded of that fact. This notion was quickly followed by her mention of a symbol on the battle flag for the Twenty-Fifth Michigan Infantry regiment. She recounted that this symbol had the words "The Battle of Tebbs Bend" arching over the image of an eagle. Her subsequent conversation recalled historical documents she reviewed from these infantry members that referred to Tebbs Bend as one of the most significant battles because of its hand-to-hand nature. She then proceeded to surprise us with accounts of how a local museum dedicated to Taylor County history had been getting more visitors since our work. Our presence had been made public when a local newspaper ran a small story on our project. She then noted that public awareness of the battlefield and people's interest in archaeology had risen.

Next we interviewed Anita Brown, landowner of the rifle pit, who also grew up on this tract of land. She was also one of the local community members who came out and engaged us about the project when we were collecting data. At that time, she had pointed questions regarding the battle, indicating she knew a good amount about its related events. During our interview Anita started off by telling us that she was happy we found something. She mentioned that she had wondered for years about the exact location of the rifle pit. Anita told us that she had grown up having picnics, riding bikes, and playing in a shallow ditch near Tebbs Bend Road where honeysuckle vines and wild roses grew. The presence of wild roses made her believe that this ditch was the location of the rifle pit; she suggested the roses were placed there as a memorial to those who died. However, her father had filled in the ditch with a bulldozer during a road-straightening project sometime while she was a teenager. Thus, she had lost direct association with this depression and it remained unknown whether it was truly related to the battle.

Anita provided us with an informative narrative about how living at Tebbs Bend had fostered an early interest in history that led to her attaining a BA in history from the University of Kentucky, before she became a nurse. Finally, she discussed how happy she was with our work and emphasized how it provided an opportunity for the history of Tebbs Bend to be passed down to younger generations. She also mentioned an increase in tourism on the Tebbs Bend ridge since the local newspaper article ran our story. She was explicit in her remarks about being proud to engage tourists when she saw them walking around near her house. Anita told us that she was both proud and happy to tell them what she knew about the battle and the position of the rifle pit in her front yard. One remark suggested that this was a new practice that started only after our work located the exact position of the rifle pit. Her new engagement with tourists indicates a stronger connection to Tebbs Bend, mediated through geophysical data and collective memory.

Concluding our interview summaries, we turn to a conversation we had with Suzanne Phillips. She and her husband, Robert Phillips, a farmer, own the cattle pen where the stockade once stood. She grew up in a house just north of the Green River Bridge, where tolls used to be collected from Civil War–era travelers on a turnpike that extended through Tebbs Bend. Yet she was well aware of the Civil War history related to the area. The Phillipses had a good idea where the stockade was positioned before we did our work, but they were happy that we confirmed its location. Suzanne mentioned that she had heard many accounts and narratives of the battle while growing up. She was happy that the stockade was properly identified on their farm, but also that stories from her childhood had been accurate. Suzanne mentioned that Robert had felt a sense of pride in learning that the stockade was on their farm. He did not grow up around the Tebbs Bend area. Therefore, he was pleased that a piece of the past was preserved on their farm. In fact, Suzanne mentioned that Robert would like to reconstruct the stockade should funding become available to do so.

Suzanne also mentioned the article in the local newspaper. She thought that since the article had appeared, more tourists were visiting Tebbs Bend and stopping at an interpretive sign positioned outside the cattle pen. The sign describes the stockade and mentions where it stood in relation to the local landform. We asked her whether she talked with tourists when they were stopped at the sign. She replied that she did not but that Robert sometimes did if he was driving by and someone was there. This difference between Anita and Suzanne probably has more to do with proximity than renewed sense of place. Anita can simply walk out her front door to engage tourists, while Suzanne would have to get in her vehicle and drive approximately half a mile down Tebbs Bend road before being able to talk with visitors.

## Discussion and Conclusion

Our interviews suggest that regardless of how people in the community formed ties to Tebbs Bend, our geophysical results provided tangible resources that they could use to negotiate their sense of place in the present. Betty Jane Gorin-Smith's intimate knowledge of primary historical records grounds her understanding of the battle and the events that unfolded on July 4, 1863. However, our results helped ground her imagination of the battle as it related to the loss of life near the Union's frontline defensive features. Moreover, it affected how she now experiences space around the battlefield. Anita Brown's experiences growing up are now reconfirmed, and a pride in the battlefield has developed that might not have been as strong prior to our work. We can see this in how she now engages tourists visiting the rifle pit. It is inspiring that she has now become a component in placemaking among tourists she encounters. Her interests are connected to her daily practices and habits; they have become an active part of how she engages with the landscape. Suzanne Phillips already had close ties to Tebbs Bend, but our work promoted a new sense of place for her husband, who did not grow up there. Our work lent a nuanced appreciation of her past and the myths and narratives about the battle that relatives and neighbors had passed down to her. Aside from the people we interviewed, Tebbs Bend is a factor in placemaking among individuals outside the immediate Tebbs Bend area, as evidenced in the rise of tourism on the ridge.

In this chapter we have argued that archaeological geophysics can provide a context whereby archaeologists can engage the public and gain their interest. Moreover, we argue that archaeological geophysics provides tangible resources that can help local agents negotiate their perception of place. This provides the opportunity for members of resident communities (*sensu* Wright 2015) to know exactly what is underfoot and to ground their imagination of the past and their connection to it in the present through these results. This allows geophysical data to be a part of the experiential process that occurs in placemaking. Wright (2015) finds that similar approaches in pubic archaeology generally lend themselves successfully to "place-attachment." Our concepts echo Wright's and aim to situate local actors in a preservation and conservation mindset based on connections to place. Therefore, we suggest that archaeological geophysics be used as more than prospection tools and data sets to explore anthropological and archaeological questions. Archaeologists should recognize how their data sets, and the narratives that are assembled from them, help structure how individuals and groups experience archaeological sites, whether they live nearby or are just visiting. As archaeologists increasingly employ geophysical methods, we hope that these aspects of the results are considered. Evaluating how geophysical results affect personal experiences at archaeological sites can certainly prove beneficial to the preservation of our past.

# 4

# Earthen Mound Construction and Composition

# 11

# The Role of Geophysics in Evaluating Structural Variation in Middle Woodland Mounds in the Lower Illinois River Valley

*Jason L. King, Duncan P. McKinnon, Jason T. Herrmann, Jane E. Buikstra, and Taylor H. Thornton*

The Illinois River Valley remains one of the most impressive and best-preserved records of Middle Woodland (2000–1550 B.P.) moundbuilding in the North American midcontinent, as affirmed by the significant number of avocational and professional excavations conducted over the last century and a half and their impact on Americanist archaeology (F. Baker et al. 1941; Buikstra 1976, 1988; Buikstra et al. 1998; D. Charles 1992, 1995; Cole and Deuel 1937; Farnsworth 2004a; McAdams 1881, 1884, 1887; Perino 1968, 1973a, 1973b, 1973c, 2006; Snyder 1895, 1909; Struever 1960, 1964; Thomas 1894). Although numerous Illinois Valley mounds have been excavated since the nineteenth century, the overall range of structural variation in Middle Woodland mounds remains only partially understood. We do not yet have a well-distributed spatial and temporal sampling of prehistoric structures at the resolution necessary to make comprehensive statements about the nature of moundbuilding, monumentalism, and the meanings encoded therein. Recent changes in archaeological ethics and laws regulating the disturbance of human graves and items of cultural patrimony, such as the Native American Graves Protection and Repatriation Act (1990) and the Illinois Human Skeletal Remains Protection Act (1989), have limited mound archaeology in Illinois and elsewhere in North America. Thus, direct inspection of mound interiors is rarely possible, necessitating new strategies for investigating structure and variability. Geophysical prospection is one noninvasive method for ethically investigating and monitoring sensitive cultural resources that would otherwise be inaccessible (Henry et al. 2014; Kassabaum et al. 2014; Monaghan and Peebles 2010; see also chapters 12 and 13 this volume).

In this chapter, we review select geophysical surveys conducted at mound sites in the Lower Illinois Valley (LIV). To contextualize our geophysical results, we begin with an overview of LIV Middle Woodland monumentalism, focusing on internal mound structure informed by archaeological investigation. We then turn to a comparative representation of proposed structural features in our geophysical data sets. Finally, we consider some issues involved in moving from methodological and empirical questions concerning the mapping of archaeological phenomena toward using geophysical data to address anthropological questions about past peoples and the landscapes they created and imbued with meaning.

## Mound Archaeology in the Illinois Valley

Mound archaeology in the Illinois Valley has its roots in the westward expansion of Euro-American settlers and their recognition that the landscapes they observed included earthworks and other built phenomena (Buikstra et al. 1998; Farnsworth 2004b; Van Gilder and Charles 2003). Avocational explorers such as William McAdams, John Francis Snyder, and John G. Henderson provided the first documented explorations of prehistoric tumuli in the region (Farnsworth 2004b and citations therein). The central question concerning mounds during the nineteenth century was who built them. Illinois Valley mounds were among those sampled to resolve this issue through Cyrus Thomas's Bureau of Ethnology explorations (Thomas 1887, 1894). While some observers noted structural details of mounds, a primary goal of these predisciplinary excavations was often the recovery of the aesthetically pleasing objects that were frequently found within the large tombs at the center of mounds.

In the early twentieth century the emerging profession of archaeology likewise focused on excavating and documenting mounds, and like their avocational predecessors, early archaeologists were often concerned with artifact recovery. Warren K. Moorehead's Illinois Valley Mounds Survey (1927–1928) was the first systematic archaeological survey of ancient monuments within the river valley. The University of Chicago Central Illinois Valley projects focused on establishing cultural units and stratigraphic sequences within Illinois and connecting them to the broader schema embodied in the Midwestern Taxonomic Method (Cole and Deuel 1937; Deuel 1935; McKern 1939).

Gregory Perino's extensive LIV fieldwork (1950–1975) constitutes the earliest efforts to systematically excavate entire prehistoric mounds and mound groups (Perino 1968, 2006). By today's standards, Perino's documentation of mound structure is limited; however, Perino recognized what are generally considered quintessential structural features of Middle Woodland mounds, such as ramp/tomb complexes and contrasting central/peripheral burial configurations. Importantly, Perino reported "loaded" mound fill composed of distinctive sediments and soils.

In 1958, the Center for American Archeology (CAA) began systematic investigation of LIV mounds when Stuart Struever initiated fieldwork at the Kamp Mound Group (McKinnon et al. 2016; Struever 1960). This initial effort in mound archaeology, coupled with Perino's work, led to an institutional focus on documenting Woodland mounds that continues today. This work was supplemented by intensive surveys of mound sites conducted in the 1970s to document the distribution of these sites along substantial transects of the valley (Buikstra 1981; D. Charles 1985; DeRousseau 1973; Holliday 1977; Palkovich 1975). Our current LIV work builds on this archaeological history and is part of a project designed to understand the manner in which monumental architecture constructed social relations within and between prehistoric communities. Of particular interest is the intersection of cosmology, ideology, ceremonialism, and monumentalism in the construction and maintenance of community at varying moments and scales of social interaction.

As an integrated component of this project, we conducted geophysical surveys at a number of LIV Middle Woodland mound sites to build a comparative corpus of geophysical signatures of mound structures (Herrmann et al. 2014; McKinnon et al. 2016). The overarching goals of these surveys include (1) addressing the *empirical* question of whether mound structures can be detected at sufficient resolution to identify mound architecture and build a body of data to evaluate regional structural variation, (2) using these data to address *anthropological* questions about the nature of Middle Woodland moundbuilding and its relation to community and cosmos, and (3) evaluating the degree to which structures remain intact and should therefore be the target of conservation efforts.

## Woodland Period Moundbuilding in the LIV

The LIV comprises the final 120 km (75 mi) of the Illinois River, its floodplain, bluffs, and associated upland features (Figure 11.1; see also Figure I.1). Prehistoric peoples undoubtedly ranged broadly across this landscape; however, their most visible signatures are hundreds of mounds built on the bluff crests and on raised geomorphic features within the valley, defined by archaeologists as bluff crest mounds and floodplain mounds, respectively (Buikstra and Charles 1999; D. Charles and Buikstra 2002; D. Charles et al. 2004; J. King et al. 2011; Ruby et al. 2005; Struever 1964). This spatial difference is thought to reflect functional and scalar dichotomies of Middle Wood-

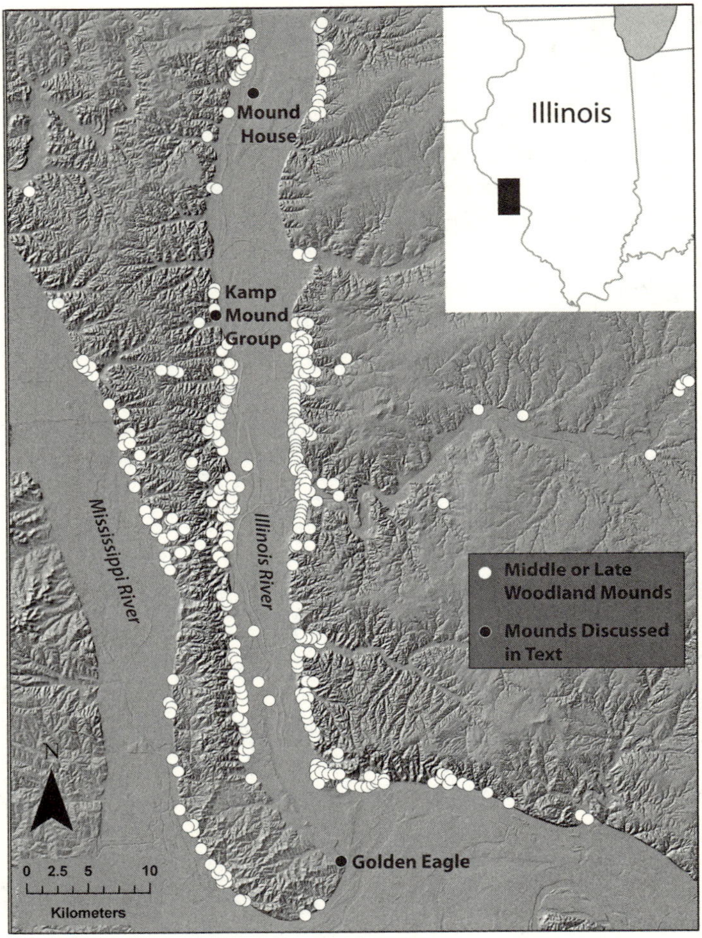


Figure 11.1. Location of sites within the Lower Illinois River Valley that are discussed in the text.

land social practices and community interactions. Bluff crest mound sites are generally considered to be residential community cemeteries. Floodplain mound groups, in contrast, have traditionally been conceptualized as supraresidential gathering spaces, anchored in part by mounds and mound-building in which funerary activity is only one of the activities reflected there. Presumably, events, funerary or otherwise, at floodplain mound groups were regularly attended and/or participated in by multiple residential units. These tumuli are frequently considered cemeteries, but despite their funerary components, neither type should be considered strictly cemeteries, but rather "platforms or stages for ceremonial performances into which some burial crypts are incorporated" (D. Charles et al. 2004:53).

Differences in the scale of activities, and presumably audience, at bluff crest and floodplain mounds appear related to the overall size and complexity of the structures. Bluff crest mounds are usually smaller than those in the floodplain, though Naples-Russell Mound 8 and Elizabeth Mounds 6 and 7 are exceptions (Buikstra and Charles 1999; Bullington 1988; D. Charles and Buikstra 2002; J. King et al. 2011; Ruby et al. 2005). Although typically smaller, bluff crest mounds generally include greater numbers of interments that are demographically representative of the communities building them (Buikstra 1976). Despite differences in use and scale, both mound types share structural similarities thought to encode commonly held meaning(s) and functions (see Hall 1979, 1997; Ruby et al. 2005; Struever and Houart 1972).

## The "Standard Model" of Middle Woodland Mounds

Archaeologists in the LIV have developed a working model of Middle Woodland mound structure anchored in Perino's observations (Buikstra and Charles 1999; Buikstra et al. 1998; Bullington 1988; D. Charles et al. 2004; Perino 1968, 2006; Van Nest 2006; Van Nest et al. 2001). This basic format includes a cleared and/or prepared surface, central tomb and ramp complex, and final capping layer or layers that effectively seal the earthen monument. The initiating event is often the creation of a cleared surface by removing all or most of the organic A horizon soil in what will be the footprint of the earliest phase of the built structure. Next, a layer or layers of light-colored sediment are deposited on this cleared surface, creating the prepared surface upon which the initial ramp and tomb complex will be built. We introduce this distinction between cleared and prepared surfaces to differentiate between removal of surficial soil and deposition of new soil before ramp and tomb construction.

Nearly all bluff crest and some floodplain Middle Woodland mortuary rituals are centered on the ramp and tomb complex. Some burials are processed through the central crypt and later interred elsewhere within the complex; some are left in the central tomb; and others are placed in pits at the periphery of the ramp. Eventually, the entire structure is buried beneath a capping layer or layers.

While the features that make up the standard model are common, variation exists (Bullington 1988; D. Charles et al. 2004). For example, the primary structures of Brangenberg Mounds 1, 2, and 3 were covered with limestone pavement (F. Baker et al. 1941; Taylor 1928). Excavation at the

Elizabeth Mounds documented several additions and modifications to the basic format, which include burning and abandonment of the central tomb prior to closing the mound with an extensive capping layer (D. Charles et al. 1988). Additionally, instances of multiple ramp and tomb structures are not uncommon. These may be superimposed, such as those interpreted in the Kamp Mound 7 geophysical data (see below) or those observed archaeologically at Pete Klunk Mound 1. They may also be horizontally related within conjoined mounds, such as examples from the Bedford, Pete Klunk, and Kamp Mound Group (McKinnon et al. 2016; Perino 1968, 2006; Struever 1960).

It is important to also distinguish between the structural and compositional properties of mounds. *Structural* elements refer to discrete building units of a mound—for example, prepared surfaces, ramp/crypts, ramp extension(s), and successive capping layers—and are basic architectural elements of mounds. *Compositional* elements, in contrast, refer to construction materials used to build structures. Julieann Van Nest and colleagues have developed a classificatory schema for Illinois Valley mound fills based on constituent materials: loaded, massive, and stratiform (Van Nest et al. 2001:634–635).

Loaded mound fills are composed of discrete depositional units with distinct boundaries, either as compositional loads or sod blocks. Compositional loading is characterized by soils from multiple sources, with contrasting soil colors and textures readily distinguishing the units, even across millennia. Sod blocks, in contrast, are discrete turf units. While their boundaries are distinct, variation observed as in situ mound fills is the result of each block crosscutting A and sub–A horizon soils. Both have a variegated appearance, but the soil color and texture variation occurs *within* distinct blocks of sod and *across* distinct compositional loads. Finally, sod blocks tend to be from a singular source. Both are generally used to build the ramps of the ramp/tomb complex (or primary mound) or structural additions.

Massive fills, in contrast, are homogeneous and have indistinct boundaries. They are typically sediments placed over the primary mound as caps or other secondary structural units.

Stratiform fills are either geogenic or anthropogenic laminated or bedded fills, such as waterlaid flood deposits or prepared sand surfaces. All three fills may be found within a single mound, although major types appear to be strongly correlated with specific structural elements. The compositional properties of these construction units may or may not create specific signatures with regard to various geophysical instruments. Detection is further complicated by the constituent soil properties of construction units, for example, sandy versus clayey soils and the many observable textural properties of soil.

## Geophysical Prospection and the Standard Model

Since 2010, we have conducted geophysical surveys at a number of Middle Woodland mound sites (see Herrmann et al. 2014; McKinnon et al. 2016 for details on data collection and processing). We focus here on results from Mound House Mound 1 (MD 1), Golden Eagle Mounds 1 and 2 (GE 1 and 2), and the Kamp Mound Group Mounds 1, 6, 7, and 10 (KP 1, 6, 7, and 10) to illustrate structural and compositional aspects as monitored by different geophysical instruments (see Figure 11.1). Results from MD 1 and GE 1 are electrical resistivity tomography (ERT) pseudosections. GE 2 was mapped using ground-penetrating radar (GPR) and magnetic gradiometry. KP 1, 6, 7, and 10 were also mapped with magnetic gradiometry. MD 1 results are informed by direct observation of archaeological soils recorded during earlier excavations at the site (Buikstra et al. 1998). The MD 1 results allow for a comparative archaeological data set when examining the ERT survey on GE 1. Previous coring at KP also permits retrospective "ground-truthing" of our geophysical results (Van Nest and Asch 2001). We organize our discussion by focusing on what we interpret as detectable structural units in stratigraphic sequence beginning with initial stages of moundbuilding.

### Cleared and/or Prepared Surfaces

ERT results from MD 1 (Figure 11.2) include a low-resistance anomaly that we interpret as a large depression related to a cleared surface at the interface of the high-resistance natural sand ridge that underlies the mound and the constructed mound ramp/tomb complex (Herrmann et al. 2014:171–172, Figure 4D). The depression is approximately 12 m wide and is centered below the archaeologically identified sod ramp (Buikstra et al. 1997; Van Nest et al. 2001). A second, smaller submound depression is present at the south end of the MD 1 ERT profile that correlates to an anthropogenic sand surface placed on the natural sand ridge (Herrmann et al. 2014:Figure 4A; Buikstra et al. 1998). There is no archaeological evidence of a prepared surface below the ramp and tomb at MD 1. A thin A horizon is reported below the excavated portion of the ramp, though it is not apparent whether it is intact (Buikstra et al. 1998). The sand surfaces reported at the south end of the mound are clearly anthropogenically prepared, though no ramp/tomb complex has been documented there. In contrast, the GE 1 ERT profile lacks such a depression (Figure 11.3). As Herrmann and colleagues (2014)

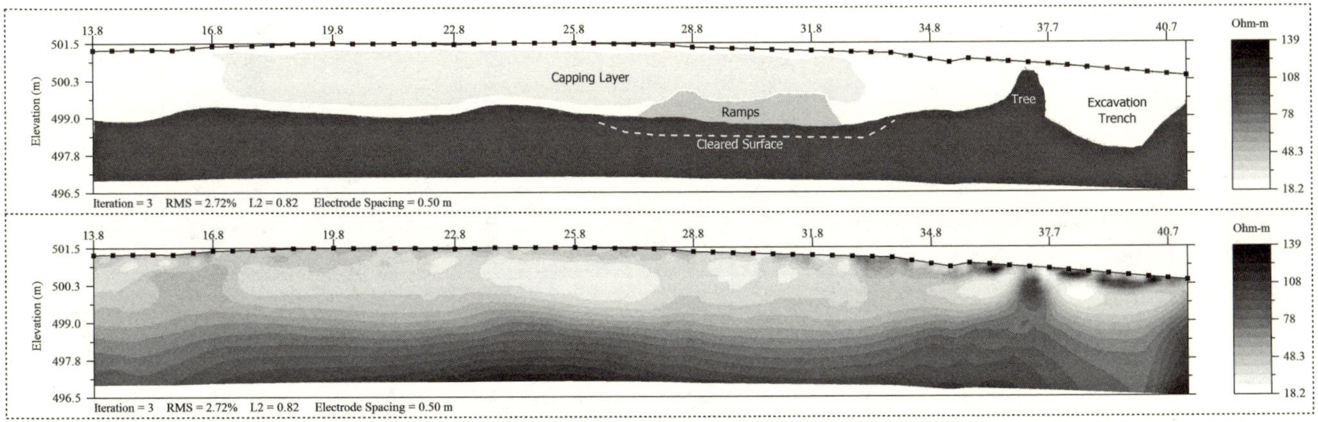


Figure 11.2. Detail of electrical resistance profile of Mound House Mound 1 with location of prepared surface and loaded fills indicated. RMS, root mean square.

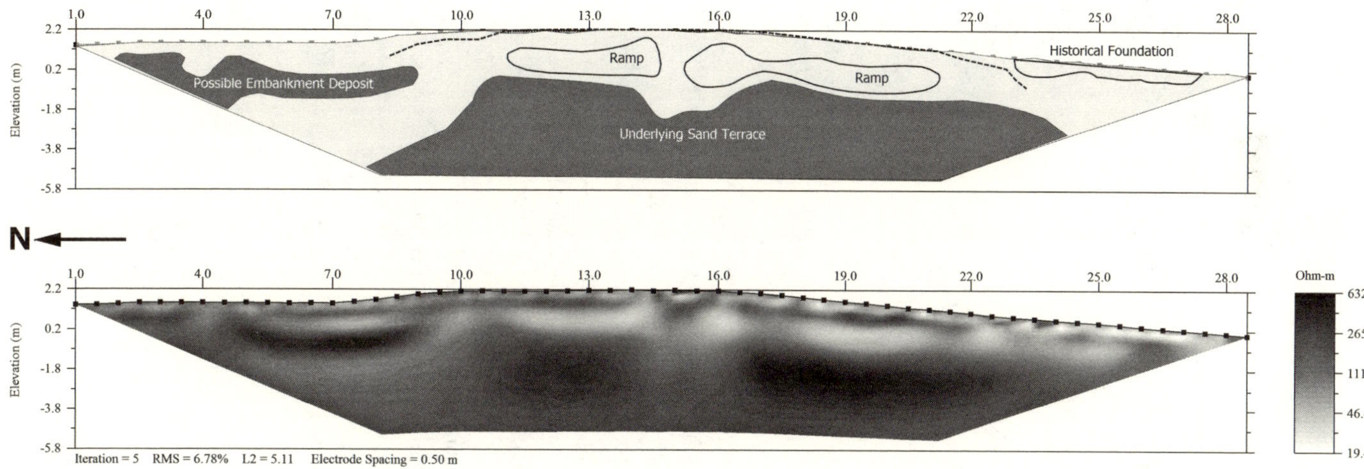


Figure 11.3. Electrical resistance profile of Golden Eagle Mound 1. RMS, root mean square.

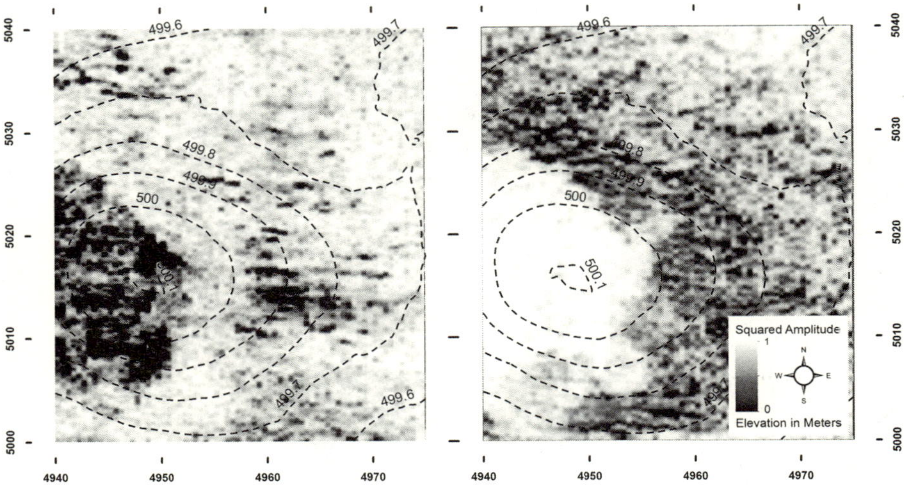


Figure 11.4. Time slices from ground-penetrating radar survey of Golden Eagle Mound 2: left, capping layer as seen in data from 9.9 to 13 ns (estimated 90 cmbs) from sensor; right, surface preparation as seen in data from 22.7 to 23.9 ns (estimated 170 cmbs) from sensor.

note, there was no expectation of detecting the 10 to 20 cm thick archaeologically observed sand layers below the southern half of MD 1, particularly when juxtaposed with the sand ridge soils on which the site was built. Similarly, we have no basis for interpreting such a structure below GE 1, though at present there is also no geophysical evidence to rule out the presence of a prepared stratiform surface either. For example, apparent submound depressions in the ERT data could represent differential water drainage patterns, a confounding factor that would directly impact resistance measurements.

Ground-penetrating radar results from GE 2 also suggest a prepared surface. Amplitude slices from GE 2 results (Figure 11.4, right) show high-amplitude reflections (27 ns; 1.8 mbs) that surround a less reflective area on which the mound is centered (Herrmann et al. 2014:Figure 6). These reflections are interpreted as the original ground surface (OGS) upon which GE 2 was built (Herrmann et al. 2014:Figure 6). The less reflective region near the center of the mound is interpreted as an approximately 15 m (~ 50 ft) area of the OGS. This area appears to be too large for a central tomb, but it is the expected location for a cleared surface.

## Ramp and Tomb Complexes

Ramp and tomb complexes are the central organizing feature of Middle Woodland tumuli. As noted, they are often constructed of distinctive soils that contrast with the underlying natural substrate and prepared stratiform surface(s). Above these complexes are the often massive, homogeneous fills that closed the tombs and form the contours of the present-day mounds. Because of contrasts between the ramps and both underlying and overlying soils, ramp and tomb structures seem to be the most easily detected and interpretable units visible in geophysical data sets. MD 1 ERT results document a feature that is consistent with a ramp and tomb complex, which is less resistant than the soil below it and more resistant than the overlying massive fills (see Figure 11.2; Herrmann et al. 2014:Figure 4). The archaeological and geophysical data suggest that the MD 1 central tomb around which the ramps were organized was not intrusive into the original ground surface. These results are important because they can be directly correlated with archaeologically observed mound structure and constituent soils. For example, the geophysical feature aligns with the sod ramp observed in Sq 81 at MD 1 (Buikstra et al. 1997).

In contrast, ERT results from GE 1 suggest a centrally placed, large, rectangular, roughly straight-walled, low-resistance depression suggestive of a central tomb that *is* intrusive into the original ground surface (Herrmann et al. 2014:Figure 5). The proposed large tomb is approximately 3 m north to south and is flanked by two low-resistance areas that are interpreted as ramps. Three plausible interpretations emerge from these data: (1) the ramp surrounds a large crypt that is several meters deep as measured from the top of the ramp to the base of the crypt; (2) there is a large crypt excavated into the original ground surface with a smaller, ramped crypt superimposed on it; or (3) there is a large crypt truncated by a looters' pit.

At GE 2, GPR high-amplitude returns (9–13 ns; 0.4 mbs) indicate the top and base of a buried structure suggestive of a primary mound (see Figure 11.4 left; Herrmann et al. 2014:Figure 6). The topographic shape of GE 2 is roughly circular; however, its subsurface amplitude geometry is strongly rectangular, which is generally unexpected for Middle Woodland structures (Herrmann et al. 2014:Figure 7). Additionally, we observe an angled, reflective interface that descends and widens in the radargram to connect the two highly reflective features in the time slices, suggesting a basic ramp form (Herrmann et al. 2014:Figure 6).

In the Kamp Mounds Group (Figure 11.5), magnetic gradiometry results from KP 1, 6, and 7 delineate apparent ramp and tomb complexes and reveal high-contrast differences between the ovoid ramp shape and the central tomb features within these mounds (McKinnon et al. 2016). Magnetic gradiometry results from KP 7 suggest the presence of a ramp and tomb complex (7A) superimposed on another (7B), although the relationship of 7A and 7B is not yet fully understood (McKinnon et al. 2016:Figure 6). Areas of high contrast could represent compositional differences, although

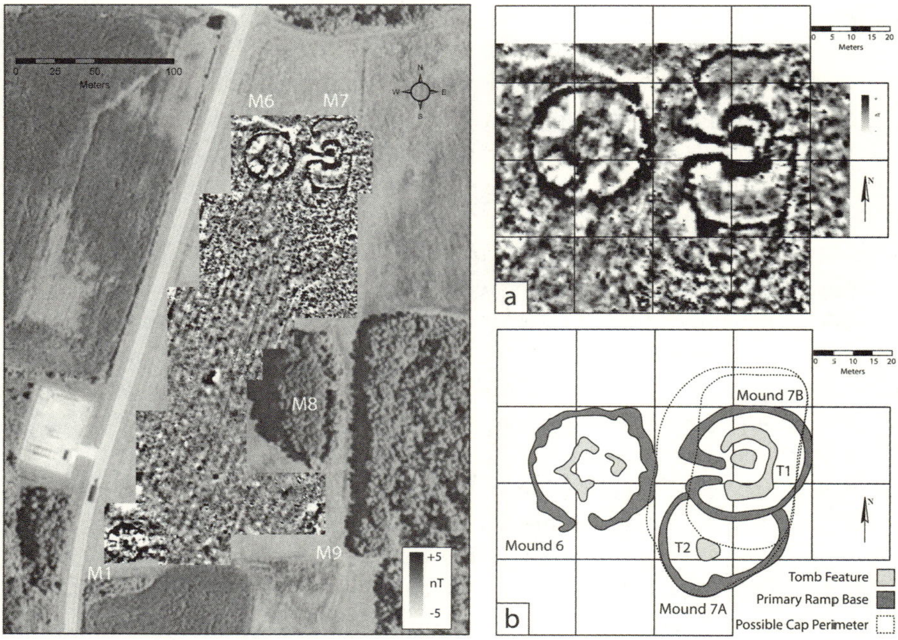


Figure 11.5. Geomagnetic results from the Kamp Mound Group: (a) results from Mounds 6 and 7; (b) interpretations from Mounds 6 and 7.

it is possible that burning on some of these structures (see D. Charles et al. 1988) could have enhanced the magnetic intensity of these features (Aspinall et al. 2008:21; Kvamme 2006b:214).

## Sod Ramps

As noted, sod blocks are single-source materials, whereas compositional loading is multisource. Thus, it is reasonable to hypothesize that the two have distinct geophysical signatures. This hypothesis has yet to be fully tested, but we have two geophysical data sets that can be supplemented by archaeological evidence. At MD 1 and KP 1, 6, and 7, mound excavations and geoarchaeological sampling have identified the presence of sod ramps (Buikstra et al. 1998; Van Nest and Asch 2001; Van Nest et al. 2001). In all four mounds, high-contrast resistance and geomagnetic signatures appear to differentiate sod ramps from surrounding structures, suggesting that it may be possible to use geophysical prospection to differentiate sod block ramps from other compositional-structural combinations. There are no direct observations of mound fills at other surveyed mounds; however, the GE 1 ERT data indicate that the hypothesized ramps are relatively low resistance compared to the overlying fill, the sediment within the tomb, and the submound surface—a pattern observed in both the MD1 ERT and excavation results. The strong contrasts in resistivity suggest distinct compositional differences between structures and may indicate sod ramps at GE 1.

### Capping Layers

Our remarks concerning capping layers are limited to three-dimensional data collected using GPR and ERT methods. Geomagnetic data provide little insight into the uppermost caps, largely owing to the inability of magnetic gradiometry to distinguish between thermoremanant and induced magnetism (see Kvamme 2006b). Capping layers are indicated primarily in their contrast to the buried architecture of the ramp and tomb complex. As Van Nest and colleagues (2001) note, capping layers and other additions to the mounds are often massive fills. That is, the soils are not (geo) archaeologically detected as loads and were not deposited as cohesive units, but rather as basket loads broadcasted over the ramp and tomb complex. In all cases, these layers are also relatively homogeneous across the structure in the geophysical data. Variation observed within them appears to be primarily the result of postconstruction disturbances, particularly historical ones as in the case of the resistance pseudosections (see Figure 11.2), or relatively nonreflective layers of soil whose profiles are dominated by historical plowing, as in GE 2.

## Geophysical Testing in the LIV: A Successful Beginning

The first stage of our geophysical testing program, detecting interpretable geophysical representations of buried prehistoric phenomena for Middle Woodland LIV sites, has been a success. Components of Middle Woodland monumental architecture are detectable via multiple sensors, and results thus far suggest that they are reasonably uniform across multiple sites. In particular, geophysical prospection at Golden Eagle and the Kamp Mound Group has allowed us to investigate internal structures we would otherwise not be able to observe. Additionally, results from these sites illustrate that significant structural remains are intact despite over a century of plowing and, in the case of Kamp (KP 1), earlier excavations (McKinnon et al. 2016:Figure 5). The ability to monitor the impact of historical modification of sites is important for cultural resource management and stewardship initiatives. Importantly, these results provide an empirical baseline for comparing LIV mounds to those in other parts of the Illinois Valley, for example, the Central Illinois Valley. Additional multisensor surveys of each mound discussed here along with additional Middle Woodland sites will only increase our resolution and understanding of structural and regional variation throughout the Illinois River Valley.

Clearly, geophysical prospection can detect buried prehistoric architecture within Middle Woodland mounds and determine, within limits, what those structures likely represent. At this level of resolution, MD 1, GE 1, and KP 1, 6, and 7 share structural details consistent with the "standard model" of Middle Woodland moundbuilding. This general uniformity suggests a shared mound-building repertoire among LIV Middle Woodland communities at floodplain sites that we have elsewhere referred to as "liturgical sequences" (Buikstra et al. 1998; J. King and Buikstra 2010). These repeated structures, which are the result of specific architectural and compositional choices made by Middle Woodland peoples, have been implicated in world renewal rituals anchored in representations of the Middle Woodland cosmos in the vertical configuration of mounds, and in re-creations of the Earth Diver's creation of land in "this world" through the use of sods (Buikstra et al. 1998; Buikstra and Charles 1999; D. Charles et al. 2004; Van Nest 2006; Van Nest et al. 2001).

That we can detect such architecture or its absence within mounds indicates that geophysical surveys of additional unexcavated mound sites will be productive and thus allow us to more firmly establish the range variation in mound configuration. With this in mind, our results thus far are encouraging. The location, frequency, and distribution of cosmologically symbolic structures and ceremonial stages are crucial for understanding the manner in which the cosmos and creation were mobilized to unite and/or differentiate various dimensions of community practically and ideologically, particularly as imagined communities who might crosscut archaeologically defined residential groups suggested by habitation sites (Bernbeck and McGuire 2011; Canuto and Yaeger 2000; Ruby et al. 2005). The ability to use geophysical data as primary data sets in future Illinois Valley research to address these questions will ensure that archaeological geophysics remains problem oriented and anthropological rather than wholly descriptive, and that the past remains peopled.

# 12
# The Anthropological Potential of Ground-Penetrating Radar for Southeastern Earthen Mound Investigations

## A Case Study from Letchworth Mounds, Tallahassee, Florida

*Daniel P. Bigman and Daniel M. Seinfeld*

The Letchworth Mounds site (8JE337) is a Middle to Late Woodland period site (A.D. 100–1000) located in the Florida Panhandle, approximately 20 km east of Tallahassee in Jefferson County (Figure 12.1; see also Figure I.1). The mounds are part of the Letchworth-Love Mounds Archaeological State Park. At approximately 16 m tall, Letchworth's Mound 1 is the tallest earthen mound in Florida. Six other mounds remain at the site, all of which are less than a meter tall. There has been relatively little scientific investigation of the mounds at Letchworth. Shovel testing, auger survey, and surface collection in the past decade have revealed some information about the site's settlement and chronology (H. Baker 2005; Glowacki 2010; Newman 1997; Tesar et al. 2003; Wheeler and Dunbar 2005). Despite this growing body of information, little is known about mound construction histories or mound functionality.

Earthen mounds are complex architectural features that are constantly affected by human-made and natural processes including looting, erosion, and tree root growth. Many earthen mounds in the southeastern United States contain burials, which limits the types of excavations that can be undertaken under state and federal law. In addition to these legal issues, the logistical costs and labor needed to conduct extensive excavations of mounded architecture can be prohibitive. Noninvasive and minimally invasive techniques, such as subsurface remote sensing, are becoming vital ways to investigate anthropological questions concerning mounds (Bigman and Lanzarone 2014; Brannan and Bigman 2014; Kassabaum et al. 2014; A. King et al. 2011; Monaghan and Peebles 2010; Pluckhahn et al. 2010). This chapter presents the results of a ground-penetrating radar (GPR) survey of five of

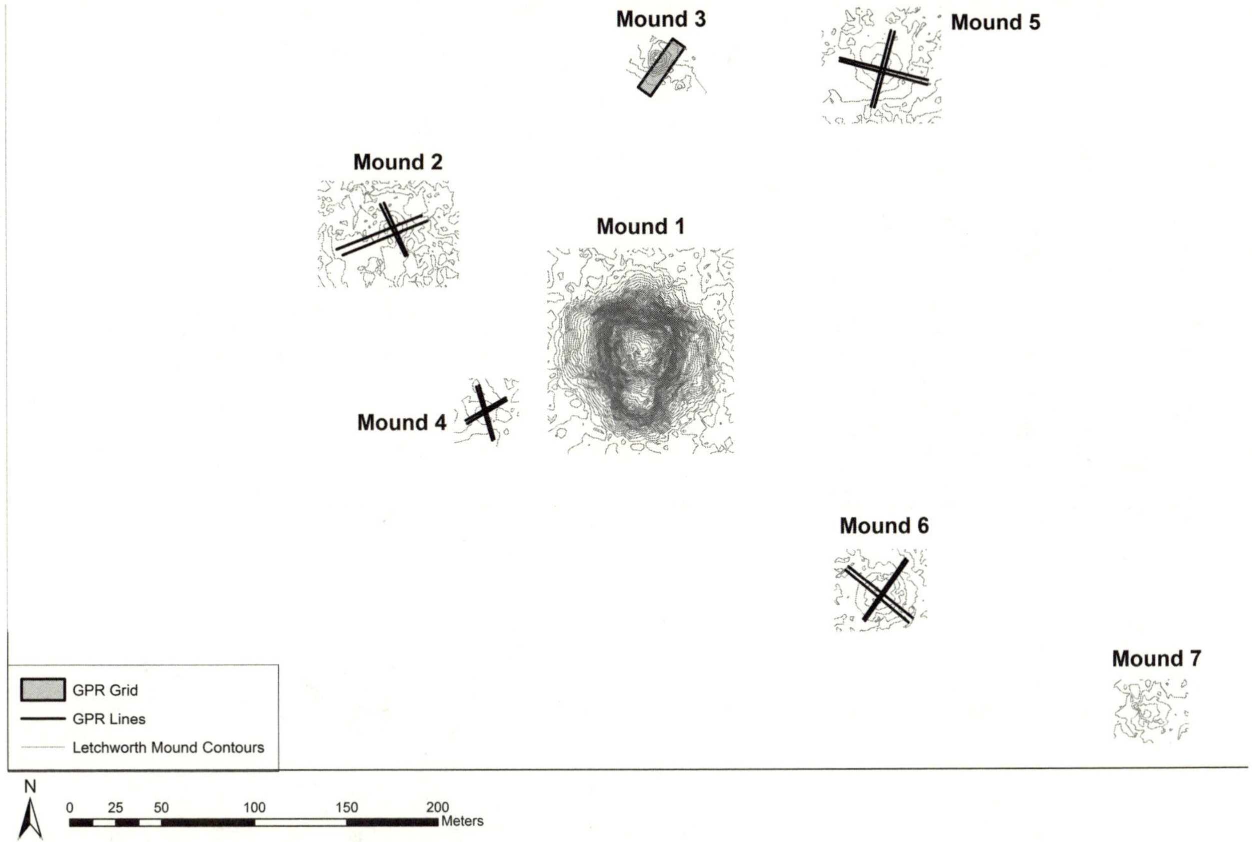


Figure 12.1. Contour map of Letchworth Mounds showing locations of GPR transects.

the six small mounds at Letchworth. These findings shed new light on the symbolic and sociopolitical significance of Letchworth's mounds in prehistory. Our study demonstrates how these minimally invasive techniques can provide substantial information about mound sites.

## Background

Previous investigations have demonstrated that the main occupation at Letchworth occurred during the Weeden Island I period (A.D. 300–700) (Newman 1997; Tesar et al. 2003), which spans the Middle and Late Woodland periods. Ongoing investigations should soon provide a more detailed understanding of the site's chronology. This period was marked by interregional interaction and ceremonial activities, often involving monumental earthen architecture (Milanich 2002; Milanich et al. 1997; Pluckhahn 2003; Wallis 2011, 2013; Willey 1949; J. M. Williams and Elliot 1998).

Middle and Late Woodland period settlement systems varied considerably. Some populations lived in small coastal camps, while others lived in large villages with monumental civic-ceremonial architecture (Milanich 1994; Pluckhahn 2003). Woodland period subsistence strategies also varied based on geography and locally available resources (Pluckhahn 2003; Ruhl 2000). Residents of some Middle and Late Woodland period sites in the southeastern United States may have practiced some form of horticulture (Gremillion 2002; Johannessen 1993; Pluckhahn 2003; Ruhl 2000; B. Smith 1989), while residents of other sites subsisted on only wild resources (Milanich et al. 1997:34; Ruhl 2000).

Letchworth is located approximately 3.5 km south of Lake Miccosukee. As with other bodies of water in north Florida, Lake Miccosukee is drained by a large sinkhole, and water levels can vary drastically (Yon 1966:19). Natural communities at Letchworth include pinelands, mixed hardwood forests, mixed pine and hardwood forests, mixed wetland forests, cypress swamp, hardwood hammocks and forest, and shrub and brush land (Florida Natural Areas Inventory 2010). This natural community contains poor agricultural soils but is ideal for hunting and fishing (Tesar et al. 2003:5). Further research on subsistence practices at the site would be needed to understand how Letchworth residents exploited the natural environment. Blanton fine sand and Pelham fine sand dominate the site, although some portions include Lucy loamy fine sand, Albany sand, and Plummer fine sand (Natural Resources Conservation Service 2013). While ground-penetrating radar (GPR) can produce useful results in a variety of geological conditions, including clay, the sandy soil at Letchworth is particularly well suited for this kind of investigation. We anticipated that signal attenuation would be slow and radar data would be clear.

## Survey Strategy

Our remote sensing work at Letchworth was designed to answer three main questions. First, were the smaller mounds constructed in a single episode or multiple episodes? Second, do any of the mounds contain subsurface features such as burials or architecture? And finally, what cultural practices and processes are reflected in the construction of the smaller mounds?

Our project conducted a GPR survey using a 400 MHz antenna on Mounds 2 through 6 (see Figure 12.1). Under sandy loam conditions devoid of clay, such as at Letchworth, the 400 MHz antenna can prospect between 3 and 5 m in depth. We collected four transects each on Mounds 2, 4, 5, and 6 because of time and resource limitations. Transects were arranged in two pairs of parallel survey lines, crossing in the approximate center of the mound to maximize coverage and efficiency. While this strategy limits the spatial extent of data recovery, we anticipate returning to Letchworth in the future to conduct a full-coverage survey of each mound with both GPR and magnetometry. We collected a survey block consisting of parallel transects at closely spaced intervals (0.5 m) on Mound 3. The justification for this strategy on Mound 3 is that the northwestern section of the mound is intercepted by a cement walkway and a large tree is located on the mound summit. The surveyor ran transects in a "snake line," collecting the first transect going southwest-northeast and alternating directions on adjacent transects. The first transect was surveyed over the peak of the mound to the east of the tree. Each additional transect was collected 0.5 m southeast of the previous transect. We surveyed a total of 16 GPR transects on Mound 3.

The GPR recorded data for 80 ns during all transects. The zero times for two-dimensional profiles were created using an automatic read-time function. We averaged the first-reflection times of all traces in the profile to arrive at the final adjusted start time. We minimally postprocessed the GPR data because they already contained a high level of clarity as raw data. A background filter was applied to all data to remove horizontal banding from profiles resulting from noise or surface objects that were the same distance away from the antennae during acquisition (Conyers 2013:123).

Following the GPR survey, Florida Bureau of Archaeological Research archaeologists conducted a series of soil augers to confirm mound stratigraphy for the small mounds. Augering complemented the geophysical investigations because the augers were also minimally invasive and relatively fast and easy to collect. We used a 2 cm diameter auger that went to a maximum depth of approximately 2.1 m when soil compaction allowed. We placed a single auger each in Mounds 2, 4, 5, and 6 in areas of the mounds that the GPR showed to be devoid of reflections indicative of burials or other archaeological features.

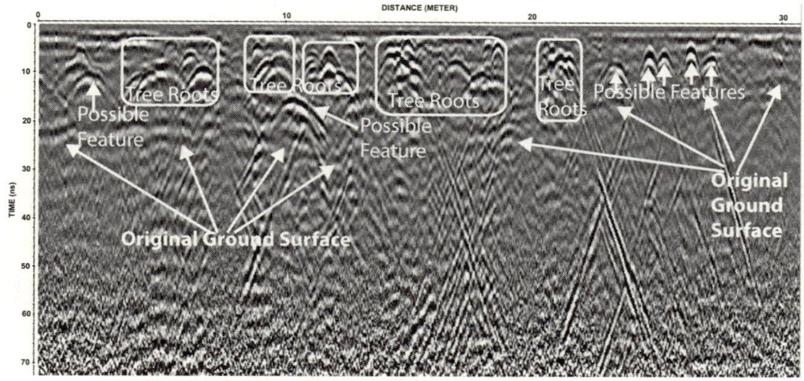


Figure 12.2. Two-dimensional profile collected on Mound 2 with a 400 MHz antenna, showing reflection events indicating possible cultural features located in the mound fill and the base of the mound.

## Results and Discussion

Only a representative profile from each mound is given in this section. Profiles were not corrected for topography because of limited time in the field. However, the clean results produced at Letchworth and the simplicity of the small mounds makes this processing step unnecessary for interpretation. The original ground surface that each mound was built on is measured as a dipping reflection. This dip indicates the greater depth the wave traveled as the antenna moved toward the mound summit; thus it took longer for the receiver to record the wave's two-way travel time. All data and interpretations are expressed in two-way travel time unless otherwise indicated.

### Mound 2

GPR data suggest that Mound 2 was constructed in a single episode. A continuous low-amplitude reflection is located between 10 and 20 ns (Figure 12.2), but some of this reflection is obscured by significant tree-root disturbance. We interpret this continuous reflection as the base surface upon which Mound 2 was constructed. Subsequent excavation of Mound 3 and soil augering suggest that this base is a sandy-clay substrate rather than a buried A horizon. Some reflection anomalies (see Figure 12.2) may represent cultural features located in the mound fill, but these reflections could be related to the tree-root disturbance. Little information can be gleaned from the GPR data collected on Mound 2 because of the extensively obscured subsurface, but no continuous reflections deeper than the one indicated suggest that this mound was constructed in a single episode.

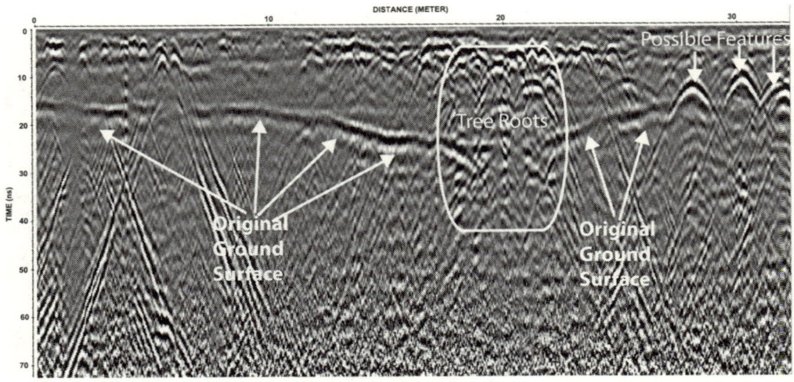


Figure 12.3. Two-dimensional profile collected on Mound 3 with a 400 MHz antenna, showing reflection events indicating probable cultural features, the base of the mound, and extensive tree-root disturbance.

The auger in Mound 2 revealed a stratigraphic profile consisting of approximately 15 cm of humic layer (10YR 4/2) over 25 cm of yellowish-brown (10YR 5/4) and 73 cm of brownish-yellow (10YR 6/6) sand fill over a thin layer (3 cm) of light gray (10YR 7/2) sand that corresponded to Stratum D. The auger showed no signs of "capping" layers.

## Mound 3

Mound 3 also appears to have been built in a single episode. A continuous reflection is observable between 15 and 25 ns (Figure 12.3). The variation in two-way travel time indicates the rise of the mound summit. Three strong hyperbolic reflections were recorded between 10 and 15 ns just north of Mound 3 (see Figure 12.3). These reflections are likely prehistoric cultural features. The GPR also recorded tree-root disturbance from a large tree located directly on top of the mound summit. The tree roots have penetrated through the single mound stage, obliterating that portion of the mound (see Figure 12.3). The disturbance terminates at approximately 40 ns, which is below the level of the original ground surface. The physical signature represented by tree roots is a series of overlapping, low- and high-amplitude, small reflections resulting from the GPR wave reflecting off small roots in various directions.

A 1 x 2 m unit excavated into Mound 3 confirmed that the mound was built in a single episode (Figure 12.4). We used the GPR data to place the unit in an area devoid of potential burials. The excavation unit in Mound 3 was largely devoid of artifacts. Stratigraphic analysis revealed that construction included digging into the substrate to a natural level of light-colored

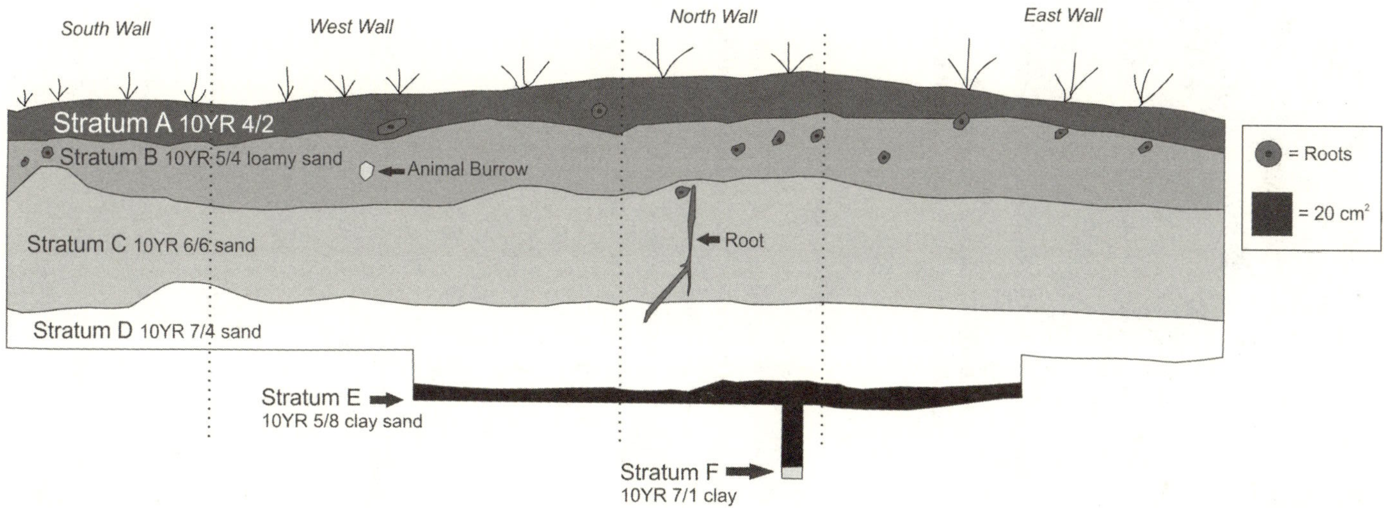


Figure 12.4. Stratigraphic map of Unit 1 in Mound 3.

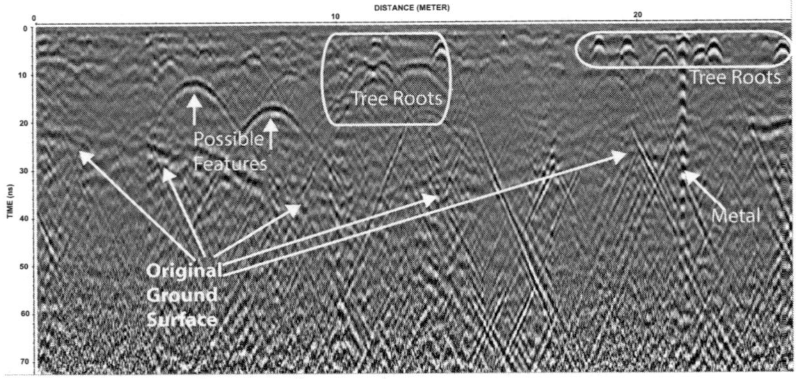


Figure 12.5. Two-dimensional profile collected on Mound 4 with a 400 MHz antenna, showing reflection events indicating probable cultural features and the base of the mound.

(10YR 7/4) sand (Stratum D) above a stratum of hard-packed natural clay (Stratum E). The Stratum D soil in Mound 3 was thinner than the Stratum D soil in a shovel test pit excavated near the unit off the mound, which suggests that the mound builders dug into this Stratum D soil during construction. The yellowish Stratum C soil was thicker in the mound than in the off-mound unit, which suggests that the mound builders sought this yellowish sand for construction. The hard-packed Stratum E is likely the reflection at the base of the mound seen in the GPR section views (see Figure 12.3).

## Mound 4

Like Mounds 2 and 3, Mound 4 has trees located on its summit; however, the tree-root disturbance appears to have impacted the subsurface minimally (Figure 12.5). Reflections from tree roots do not reach the original ground surface, which is shown as a continuous dipping reflection (see Figure 12.5). The absence of any other continuous reflections suggests Mound 4 was constructed in a single episode, and the hyperbolic reflections indicate probable cultural features or material culture in the mound fill (see Figure 12.5). These do not appear to be intrusive events following completion of the mound since there are no reflections suggestive of pit walls beginning at the ground surface. Furthermore, there is no break in the high-amplitude ground-surface reflection (i.e., a low-amplitude ground-surface reflection over a hyperbolic reflection) indicative of a burial shaft or looters' pit. The location and geometry of these reflections suggest that the features may have been placed in the mound during construction.

The soil auger in Mound 4 showed that, like the other mounds, it was

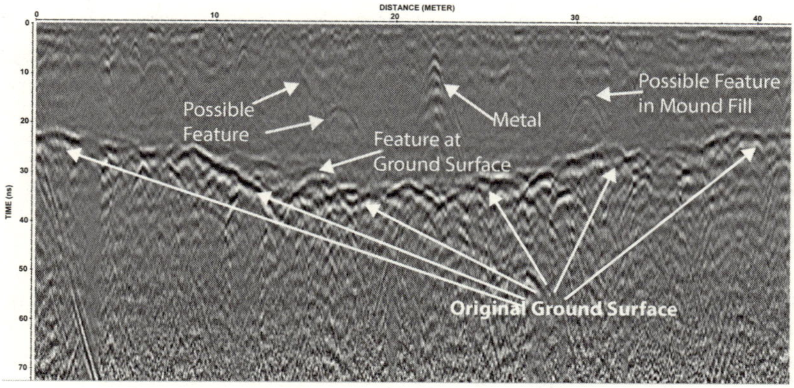


Figure 12.6. Two-dimensional profile collected on Mound 5 with a 400 MHz antenna, showing reflection events indicating possible cultural features at various depths and the base of the mound.

constructed from approximately a meter of yellowish sand (25 cm of 10YR 5/6 sand and 80 cm of 10YR 6/6 sand). The auger mirrored the GPR results in that there was evidence of only a single construction episode.

## Mound 5

Mound 5 contains geophysical evidence for a variety of possible cultural features at different depths (Figure 12.6). Two different dipping reflections beginning at the ground surface and connecting at approximately 18 ns may represent the sides of a pit. Just below this, beginning at 19 ns, is a hyperbolic reflection marking the probable bottom of the interpreted pit (see Figure 12.6). Models of GPR reflections for excavated pits have resulted in similar reflection geometries (Conyers 2013). These findings suggest that a pit was excavated, or a burial interred, in Mound 5 following its construction. A small low-amplitude reflection is located at 12 ns, which may represent cultural activity (see Figure 12.6). This reflection is too small to definitively identify and may simply represent a reflection of an artifact or increased heterogeneity in the mound fill. The GPR identified another possible cultural feature at 30 ns, which is the approximate depth of the original ground surface (see Figure 12.6). This correlation of depth suggests premound construction activity. Possibly, it represents a burial or artifacts deposited on the ground and covered, or a pit dug into the original ground surface and filled with a contrasting material such as organic refuse. The reflection anomalies identified at varying depths and with varying signatures may indicate complicated processes related to mound construction. A strictly prescribed recipe or protocol may have been followed to sanctify this monument. Each step (premound construction, mound construction, and postmound construction) may have been imbued with shared meaning, and the fulfillment of each step might have been required before transitioning to the next phase of construction. Again, the absence of a second continuous reflection indicates that Mound 5 was built in a single episode.

Soil augers of Mound 5 contained approximately 120 cm of brownish-yellow (10YR 6/6) sand over 35 cm of very pale brown (10YR 7/4 and 10YR 7/3) sand and Stratum D (10YR 8/1) sand. Much like Mound 3, Mounds 2, 4, and 5 seem to have been constructed by digging a hole, placing deposits in the ground, and building a mound from the yellowish-brown Stratum C soil.

## Mound 6

Mound 6 is the only small mound that we studied at Letchworth with evidence for multiple construction stages. A continuous reflection begins at 25 ns and dips to 40 ns at the mound peak (Figure 12.7). The GPR recorded a second continuous reflection below this layer, beginning at approximately 40 ns on the left-hand (western) side of the profile, which dips to below 60 ns in the center of the profile (see Figure 12.7). The two anomalies are somewhat parallel, but the time difference between them is greater in the center of the transect (from the current mound peak) compared to the beginning or end. This pattern implies that the shallower anomaly is a raised surface. We interpret this surface as a previous mound peak that was expanded on or refurbished after some period of use. We interpret the deeper anomaly as the original ground surface on which Mound 6 was constructed. GPR data from Mound 6 also indicate several possible features located in the mound fill that may have been interred during mound construction, and two possible interments of caches or burials excavated into the final mound peak following construction (see Figure 12.7). Those identified here exhibit a low-amplitude reflection at the ground surface, suggesting looser soils containing more pores filled with air (less contrast in dielectric constant with the layer of air above the mound). This signature is typical of grave shafts where excavated pits are filled in with less compact soil from the same host matrix (Bigman 2014; Conyers 2013). The hyperbolic reflections recorded at approximately 12 ns are located just below these low-amplitude breaks in the ground surface reflection, indicating the wave is reflected from the pit bottom. These two anomalies may represent more recent activity, such as looting, but there was no observable looting scar on the surface. The lack of a looting scar is in contrast to the open looters' pit located on the summit of Mound 1. The specific signatures of these two reflective targets coupled with limited evidence suggestive of looters' pits lead us to believe these anomalies were deposited in prehistory.

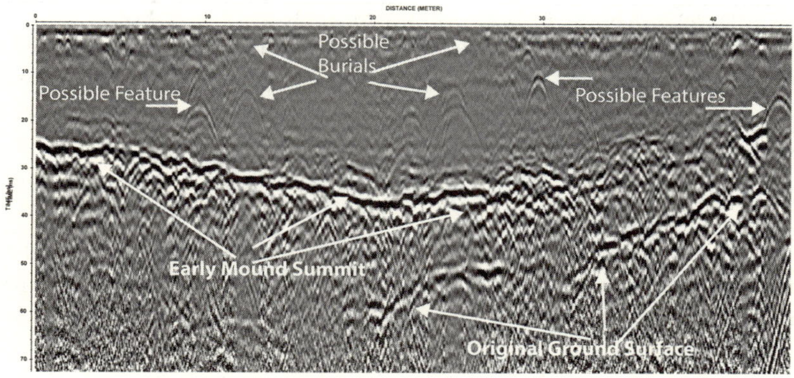


Figure 12.7. Two-dimensional profile collected on Mound 6 with a 400 MHz antenna, showing reflection events indicating possible cultural features in mound fill and at the mound summit, the summit of the first mound stage, and the base of the mound.

Materials reportedly found from earlier ground disturbance suggest that the anomalies in the Mound 6 GPR profiles may be burials. Bone fragments have been found eroding from the surface and were reburied in a shovel test (Glowacki 2010; Wheeler and Dunbar 2005:3). Louis Tesar (personal communication 2013) noted that in the early 2000s, a Florida Fish and Wildlife officer found exposed mica and pottery when a private landowner cleared pine trees from the surface of Mound 6. Such items are commonly found in Swift Creek and Weeden Island burials, such as at the McKeithen mounds (Milanich et al. 1997) and the Block Sterns site (Jones et al. 1998).

Augers confirmed that Mound 6 was distinct from the other, smaller mounds at Letchworth. As the GPR profile indicated, Mound 6 was constructed in two distinct phases with different sands. The lower level consists of yellowish (10YR 6/6) sand and the upper portion is made from a pale brown (10YR 6/3) sand. A shallow lens containing a high concentration of charred wood separates the two levels. It is unclear whether the soil auger may have gone through a burned feature.

## Conclusion

The smaller mounds at Letchworth are likely sand mounds similar to Middle and Late Woodland period communal mounds found on the Gulf Coast (Russo et al. 2014; Russo et al. 2009). GPR, augering, and excavation show that apart from Mound 6, these mounds were built in single episodes. Based on comparison with contemporary mounds in northern Florida (Jones et al. 1998; Milanich et al. 1997; Russo et al. 2014; Wallis 2013; Willey 1949), reflective targets found in Mounds 2–6 may represent burials, structural remains, and ritual deposits containing pottery and other items. At Letchworth, fragments of human bone have been found eroding from Mound 6, suggesting a possible mortuary deposit in at least one of the mounds (Glowacki 2010).

Our investigations reveal a scenario for the construction processes of the small mounds at Letchworth. These mounds were likely each built in a single episode in which the mound builders may have excavated into the soil until they reached a level of light-colored sand. Based on the GPR profiles, burials or other types of deposits were made on the white sand near the base of the mound. The hole was then filled and a mound formed using sand that was likely gathered from near the mounds. In contrast, the largest mound at the site (Mound 1) appears to have been constructed on top of the natural ground surface (Seinfeld and Bigman 2013). Mound 6 seems to be unique among the smaller mounds in that it may have been constructed in two phases.

As Knight (1986) discussed, southeastern Native American cultures assigned great significance to the communal ritualized act of mound construction. During the Weeden Island I period, mounds tended to be constructed quickly and featured sealed deposits of exotic goods (Milanich et al. 1997; Sears 1958; Wallis 2013). Wallis (2013:216) described such mounds as exclusive events that sealed a moment in history in a way that could not be "reworked or reappropriated." The reflections found in the GPR may be such deposits.

Making deposits in the light-colored Stratum D sand and then constructing the mound may have been symbolically significant to the mound builders at Letchworth. Color symbolism was employed at Letchworth's Mound 1. Each of the four cardinally oriented platforms on Mound 1 is constructed from a different-colored sand (Seinfeld and Bigman 2013). Color symbolism and directionality are of great significance to Native American cosmologies (DeBoer 2005). Placing burials or other deposits at the base of the mounds in light-colored strata may have been an enactment of the group's cosmology through moundbuilding, much as Kidder (2010, 2011) has suggested for the mounds at Poverty Point. Such use of architecture to express sacred and cosmological themes is a common feature around the world, from South Asian temples to Renaissance cathedrals (Rapoport 1982:27–28).

At a broad level, the mound complex at Letchworth represents the construction of a sacred landscape that helped integrate different groups into a community (see Pluckhahn 2010). Constructed features formed a complex for ritual and interaction at Letchworth. The smaller mounds (Mounds

2–7) are similar to contemporary village mounds found elsewhere in north Florida (Russo et al. 2014; Russo et al. 2009; Wallis 2013). GPR evidence of unexcavated deposits compared to other contemporary sites (Milanich 2002; Milanich et al. 1997; Wallis 2013) suggests that people made ritual deposits in these mounds, possibly including burials. Constructing several of these mounds around the monumental Mound 1 may represent a means of integrating diverse subgroups within a single location through shared ritual experiences involving mounded architecture (see Kujit 1996; Pluckhahn 2010).

We were able to learn new information about the mounds at Letchworth using GPR, soil augering, and small-scale excavation. These minimally invasive techniques caused little or no damage to the mounds and they were completed in a relatively short time frame for little cost. Ongoing research at Letchworth will continue to employ subsurface remote sensing to gain insight into a more extensive area of the site as well as more intensive investigations into specific mounds. The GPR data will also help guide investigations into the chronology of the mounds and how they fit into the cultural landscape of Woodland period northern Florida.

## Acknowledgments

Special thanks to the Southeast Archaeological Center (SEAC), especially Mike Seibert, Jessica McNeil, John Cornelison, and David Morgan for working with us in using the 400 Mhz GPR. Also thanks to Mike Russo of SEAC for his insight and comments. We also thank our volunteers from the Panhandle Archaeological Society at Tallahassee, John Grant, Ed Green, and Daniel Vasquez. We also thank our student volunteers Grant Stauffer and Lela Boswell. This work was made possible by the support of the Bureau of Archaeological Research and the Florida Park Service. Finally, we would like to thank Duncan McKinnon and Bryan Haley for their encouragement and patience throughout this process.

# 13
# Exploring the Deepest Reaches of Arkansas's Tallest Mounds with Electrical Resistivity Tomography

*James Zimmer-Dauphinee*

Remote sensing in archaeology is often used as merely an effective means to identify the best locations to place excavations. While it is certainly useful for this purpose, there is a rapidly growing consensus that archaeological remote sensing is, beyond anomaly detection, capable of both generating and addressing archaeological questions directly, as is demonstrated throughout this volume. The ability to rapidly survey increasingly large areas opens the door to examining settlement and activity patterns on a scale that would have been totally inaccessible from an excavation perspective. This makes it possible to ask and occasionally even answer questions about the structure, organization, and activities of entire settlements.

Furthermore, remote sensing techniques allow researchers to pose and address questions about sites that may be too culturally or historically important to examine through destructive excavation. Nevertheless, when the features of interest are buried very deeply, by either natural or anthropogenic processes, geophysical methods commonly used in archaeology fail to reach the target depth. This is especially true in geologic environments such as those in much of the mid-South and Southeast, where soils with high clay content and therefore low resistance rapidly attenuate the signal of ground-penetrating radar (GPR), the one method that in proper conditions may approach deeply buried stratigraphy (Conyers 2013). In such locations, if archaeological geophysical methods are to be utilized, new methodologies of data collection and analysis such as electrical resistivity tomography (ERT) become necessary.

## Toltec Mounds

Located approximately 24 km southeast of Little Rock, Arkansas, Toltec Mounds Archeological State Park (3LN42) is one of the largest mound centers in the region and is home to two of the tallest mounds in Arkansas (see Figure I.1). Constructed by people of the Plum Bayou Culture between 700 and 1000 C.E., 18 mounds are arranged around two plazas (Rolingson 1982:198). The complex, spanning approximately 42 ha, is bounded on three sides by an approximately 1.6 km long ditch and embankment (Rolingson 2012). The fourth side is bounded by Mound Lake, a cutoff of the Arkansas River. Much of the site has been severely damaged or destroyed by more than a century of cultivation, with only three mounds and portions of the embankment remaining readily visible today (Rolingson 2012). Nevertheless, the tallest mound, the conical Mound A, still reaches a height of approximately 15 m, and Mound B, a rectangular platform mound, remains 11.5 m tall.

The tremendous scale of the site is especially remarkable in that the complex was constructed and used primarily between 700 and 1000 C.E. during the Terminal Woodland period (Rolingson 2012). This is generally thought of as a time of cultural decline with evidence of a decrease in mortuary ritual complexity, greatly simplified ceramics, and few exotic trade goods (Nassaney 1994). While the ceramics found at Toltec indeed have little decoration and there is little evidence of the broad-scale foreign exchange that is characteristic of the earlier Middle Woodland or later Mississippian periods, the presence of such a large ceremonial center suggests a level of organization and collective effort that do not match the narrative of a society in decline.

The organization of Toltec's mounds around two plazas is equally unexpected. Woodland mounds are usually conical and the complexes have little overall organization. In contrast, Mississippian mound complexes are better known for containing platform mounds arranged around central plazas (Rolingson 1982). Toltec's careful organization and plazas are therefore more typical of Mississippian than of Woodland complexes. Even among similarly organized ceremonial complexes, Toltec is unusual, as in such instances the tallest mound is almost always rectangular or square with a large platform at the top (Rolingson 1982). Mound A is shaped like an elongated cone with a small flat surface at the top, while Mound B has a very distinctive rectangular shape and large platform surface. This was confirmed by the high-resolution elevation model generated during the course of this research (Figure 13.1). It has been proposed that Mound A was once a platform mound, but severe damage from erosion may have reshaped it to its current conical form. Similarly, it is plausible to imagine that the rectangular shape of Mound B may be the result of later modifications during the Mississippian period, or even systematic damage from cultivation and erosion; however, before this research, none of these hypotheses had been tested (Rolingson 2012).

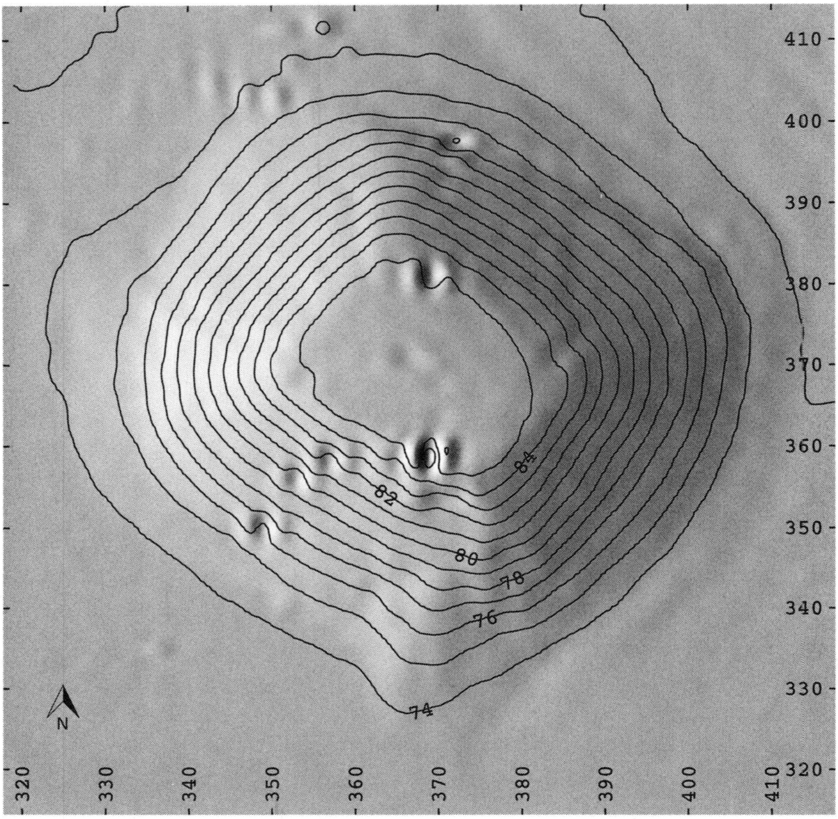


Figure 13.1. High-resolution elevation data confirm the rectangular shape of Mound B.

Clearly, developing an understanding of the two most prominent features at Toltec is crucial for understanding the site as a whole and in turn for examining the ways in which Toltec Mounds challenge the standard conception of the Terminal Woodland period in this region. Yet hardly any research has been completed on either Mound A or Mound B. Some small clues about the internal structure of the mounds can be found in the journal of Edward Palmer, who in 1883 explored the site as part of his mound survey for the Bureau of Ethnology (Jeter 2009). Palmer excavated trenches in five of the mounds, including trenches 3 m (10 ft) deep in the top of both Mound A and Mound B, and briefly described in his notes the stratigraphy of each excavation. This is one of the only sources of information about the stratigraphy of these mounds; however, it must be considered with caution. Palmer's initial description of the size and shape of the mounds reported

Mound A as being 30 m (100 ft) tall and nearly square, while Mound B was supposedly 23 m (75 ft) tall. This account was so inaccurate that Cyrus Thomas required him to return to the complex and repeat the survey in 1884 (Jeter 2009). Palmer's second survey apparently fared little better, as Thomas, in 1890, sent William H. Holmes to repeat the survey a third time, and this effort is the one that was used in the final report (McMullan 2001). This casts doubt on the validity of Palmer's notes and descriptions of the stratigraphy, but they do provide tantalizing clues.

Nearly one hundred years later, the Arkansas Archeological Survey and Arkansas State Parks gained control of the site. While much excellent research at Toltec has been completed since this time, the focus regarding Mounds A and B has been preservation. The only detailed description of the stratigraphy of Mound B comes from an effort in 1979 to stabilize an erosional gully on the southernmost corner of the mound. As part of this stabilization project, an 11 m profile was cut on the western edge of the gully, revealing some of the outer stratigraphy of the mound (Rolingson 1982). Otherwise, the mounds have been left essentially undisturbed by scientific study. Indeed, Martha Rolingson, survey station archaeologist at the park for nearly 30 years, characterized the limited data on the largest mounds as one of the "biggest gaps in knowledge" about Toltec and stated that this "lack of information severely hampers our overall understanding of the activities on the site" (Rolingson 2012:3).

Resolving this lack of data is not a trivial problem. When the park was first founded in 1975, excavations on Mounds A and B were avoided because of the great cost such a project would incur. It was therefore decided that the most effective use of limited resources would be to focus efforts on other parts of the site (Rolingson 2012). Today limited resources remain a problem. Mounds A and B are so large that a full-scale excavation would be prohibitively expensive. However, other considerations render such a project untenable. The cultural, historical, and archaeological importance of the mounds and an interest in their preservation make the destructive process of excavation an unreasonable method of data collection.

Furthermore, an examination of the surface of the mounds is insufficient because the relationship between the current surface and previous morphologies is not immediately evident. After approximately 1000 C.E. the site was abandoned and was not reoccupied until later, during the Mississippian sequence, by cultures who do not seem to have had any direct relation to the original builders of the mounds (Rolingson 2002:56). It is possible that significant modifications to the mounds may have occurred during this time, changing the morphology significantly.

The surrounding region was divided among speculators and sold following the removal of the Quapaw and French residents from the region in 1833. By the middle of the nineteenth century the land that would become known as Toltec had been purchased as a plantation, and over approximately the next 125 years it was cultivated for cotton, soy, and rice and used as pasture land (McMullan 2001). This was a very destructive time for the site, as entire mounds were used to fill borrow pits and level road grades. Erosion also likely played a large role in damaging the site, as letters from landowner Mary Eliza Knapp before the twentieth century describe artifacts and remains eroding out of the mounds (Rolingson 1982). Destruction only accelerated with intensive agriculture. Houses were built on the land, and rice levies were constructed on the northern half of the site. Differential land uses caused differential damage on the northern and southern portions of the site, and while all but three of the mounds were destroyed and much of the embankment was leveled, it is unclear exactly what effects these activities may have had on the integrity of the remaining mounds (Rolingson 2012). Without an evaluation of the effects of nearly a thousand years of natural and anthropogenic modification to the mounds it is difficult to ask, much less reasonably answer, questions about their construction, use, and meaning.

## Geophysical Methods

The vital need for more data and the impossibility of gathering the needed data from excavations or visual observation highlight the necessity of using geophysical techniques. Unfortunately, traditional archaeological geophysics methods are also lacking. Magnetometry and conductivity provide only two-dimensional representations of near-surface deposits and thus are incapable of examining questions regarding the construction and modification of the mounds. Both depth and three-dimensional modeling, or at a minimum vertical profiles, are required to answer those questions. Ground-penetrating radar is capable of producing profiles and a fully three-dimensional model of the subsurface that may be capable of examining construction layers and modifications of Mound C, which stands only a few meters tall. However, in the highly conductive clay soils at Toltec the signal can produce returns to a depth of only around 3 m, even with the use of a 200 MHz antenna. For mounds as large as A and B, 3 m can address only the end of mound construction and cannot capture deeper features or evidence of earlier mound morphologies.

The remaining common archaeological geophysics method is electrical resistance or resistivity. Standard archaeological configurations of resistivity meters utilize between four and eight electrodes, with the depth of investigation controlled by the spacing of the electrodes. Therefore, simply increasing the distance between electrodes may increase the depth recorded. Frames usually employed in archaeological surveys can allow for measurements of depths up to 2 m; however, geologic applications of resistivity often utilize long cables to separate probes and may record measurements at depths of hundreds of meters. Furthermore, an automated multiprobe array utilizing as many as 96 or even more electrodes with a multiplexer can collect data from many depths and locations at once, allowing the rapid collection of large amounts of data. Resistivity surveys are therefore capable of collecting data to arbitrary depths, and the depth of investigation can be controlled for each measurement. In conjunction with mathematical modeling, this allows for the construction of profiles, which can be combined to approximate three-dimensional reconstructions of the subsurface resistivity distribution of an entire mound. Though these reconstructions are lower resolution than may be achieved by GPR, the resulting models are sufficient to begin to examine changes in mound morphology.

## Electrical Resistivity Tomography

A detailed discussion of the inner workings of ERT may be found in the excellent work by Tsourlos (1995); however, a brief discussion here will allow the reader to understand the differences between more traditional archaeology-based resistance mapping and ERT models.

Many archaeologists use the terms "resistance" and "resistivity" nearly interchangeably, but there is a significant difference between the physical meanings of the two words. Electrical resistance is not a property of the material being measured; rather, it is a function of both the material and the amount of material through which the electricity passes. For instance, the resistance of a small piece of copper wire is much less than that of a small piece of iron. However, the resistance of a thousand kilometers of copper wire will be much greater than that of the small piece of iron. Resistivity, in constrast, may be thought of as resistance normalized by the amount of material through which the electricity passes and is a physical property of the material measured. It is the inverse of conductivity, the measure of how easy it is for electricity to pass through a given amount of material. Therefore, if the current must, for example, pass through more of a homogeneous soil, the resistance will change but the resistivity will not.

The implication of this distinction is that when the spacing of electrodes is changed to measure the resistance at multiple depths, the amount of material through which the current is passing will also change, making a direct comparison of resistance between different depths impossible. This is not commonly an issue in archaeological applications, as the spacing of electrodes for a single depth remains consistent and so the resistance values are comparable. This is quite effective for identifying horizontal variations in resistance that may result from walls, ditches, or other archaeological features but is not capable of examining vertical variations resulting from multiple construction phases and later modifications.

To answer these archaeological questions, it is necessary to build profiles and three-dimensional models of the subsurface. This requires mapping not the variation in measured resistance but rather the variation in resistivity. Unfortunately, it is impossible to directly measure the resistivity of the subsurface from the ground surface. It is also impossible to directly calculate the resistivity of the subsurface, as the exact path the electricity will take through the soil will be changed by any inhomogeneity within the soil. This path cannot be calculated without first knowing the true resistivity distribution of the subsurface, which is precisely the variable the researcher is trying to discover.

Fortunately, there is a way of resolving this dilemma computationally, known as computed tomography or "inversion." To begin, an estimate is made of the true resistivity distribution of the subsurface. Computer software then uses this estimate to calculate the measurements that would have been recorded on the surface had the ground matched the estimation. The estimation can then be edited to make the calculated measurements more closely match the true measurements recorded in the field. Another set of hypothetical measurements are then calculated using the new estimate of the resistivity distribution, which can then be further refined. This process is repeated until the model stabilizes and the calculated measurements closely match the recorded ones, thereby providing an accurate model of the subsurface. In this way, data collected along a transect can be processed in two dimensions to generate a profile.

It is also possible to generate three-dimensional models of the subsurface resistivity distribution. Theoretically, placing all of the probes for the survey and then recording measurements between all independent combinations of electrodes could gain the maximum amount of information and therefore the highest-resolution results. In practice, to survey Mound B at Toltec in this way with the same electrode positions used for this study would have required 2,430 electrodes and could have recorded over 4.3 trillion measurements. To provide some context for this scale, if each measurement were recorded as only one byte of data, it would take over four terabytes to simply store the measurements. Collecting and processing data on this scale are impractical with current technology. As an alternative, a series of dense parallel transects may be combined and processed in three dimensions to generate a quasi three-dimensional model.

## Applications at Toltec

An ERT survey was completed at Toltec using a Syscal Pro Switch 96 multichannel resistivity instrument, rented for one week in March 2013 through the support of Arkansas State Parks and the Arkansas Archeological Survey. Limitations of time and money made it necessary to choose either Mound A or Mound B as a primary area of interest. Though little is known about either mound, the repair of the erosional gully on the southern corner of Mound B offered some clues regarding the internal structure of the mound that could be compared to the survey results. For this reason, Mound B was selected as the primary study area and Mound A became a secondary area of interest to be surveyed as time permitted.

Forty-nine transects were used in the survey of Mound B, for a total of 2,430 electrode insertions and more than 60,100 measurements. Transects were arranged on the north-south axis and spaced at 2 m intervals. Each transect began at the same southern coordinate and extended 94 to 100 m to the north, with electrodes placed every 2 m along the surface of the mound. Once the survey of Mound B was completed, the remaining time was utilized to survey 15 transects across Mound A. These transects were oriented in relation to the lakeshore adjacent to Mound A. The limited time frame did not allow for the completion of as many transects as on Mound B. Therefore, seven transects spaced at 15 m intervals were placed perpendicular to the lakeshore, and eight were placed parallel to the lakeshore, six separated by 10 m and two by 20 m, covering the mound as completely as possible.

The data sets resulting from each transect were cleaned and topologically corrected using the Prosys II software and processed using the RES2DINV software, resulting in two-dimensional profiles of the resistivity distribution. The data from Mound B were also processed in RES3DINV to generate a three-dimensional model. Three-dimensional inversion requires that the spacing between transects be not more than twice the electrode spacing (Papadopoulos and Tsourlos 2007). Therefore, the data from Mound A, at 7 to 10 times the electrode spacing, could not be combined into a three-dimensional model but had to be examined as individual profiles. Finally, the data were visualized, analyzed, and mapped in the open-source software ParaView, GRASS GIS, and QGIS, which allowed the researcher to slice, threshold, and otherwise manipulate the volume and profile data sets (GRASS Development Team 2012; Henderson 2004; QGIS Development Team 2009).

## Results

The profiles and three-dimensional model of Mound B reveal a three-layer model of soil stratigraphy (Figure 13.2). Approximately 8–10 m below the ground surface is a very high-resistivity base layer. Fresh water is quite resistant to the flow of electricity, and it is likely that this layer represents the water table. The middle layer is composed of a highly conductive material. It is likely that this stage is associated with clay deposits, which are typically very conductive. Away from the mound, this layer is consistently between 6 and 10 m thick. Finally, the surface layer of the model is highly resistant and dynamic, contrasting sharply with the relatively homogeneous middle layer and perhaps indicating anthropogenic disturbance. For the several transects recorded off the mound, this layer appears to be quite thin, at most 1 or 2 m thick (see Figure 13.2). In some places it nearly disappears entirely. As the survey was completed with electrodes spaced every 2 m, the maximum vertical resolution that may be expected is approximately 1 m. It is possible that this highly disturbed layer appears thicker than it is in reality because of the relatively low resolution.

The same basic model remains in regions covered by Mound B (Figure 13.3). The base layer is still evident in many of the profiles, though as the mound rises, the lowest measured depth also increases, bringing it above the water table in some places. The middle layer also persists and makes up much of the mound itself, becoming in some places at least 13 m thick. This indicates that much of the mound's internal structure is composed of the same clay material found below the ground surface, rather than primarily of midden deposits, which would likely yield high resistivity measurements. This layer remains, for the most part, quite homogeneous. This homogeneity may at least in part be the result of a decrease in resolution that occurs as depth increases. Any small anomalies that might be caused by deposits of other materials may be too small to register, giving a homogeneous appearance. There do appear to be some intrusions of high-resistivity material into the lower stage of the mound. Two intrusions, visible near the center of the mound platform, persist across several transects (see Figure 13.3). Thresholding the three-dimensional model to view only regions with a resistivity below 60 ohm-meters reveals that these intrusions appear to form two trenches, oriented roughly east-west, in the top of the low-resistivity layer

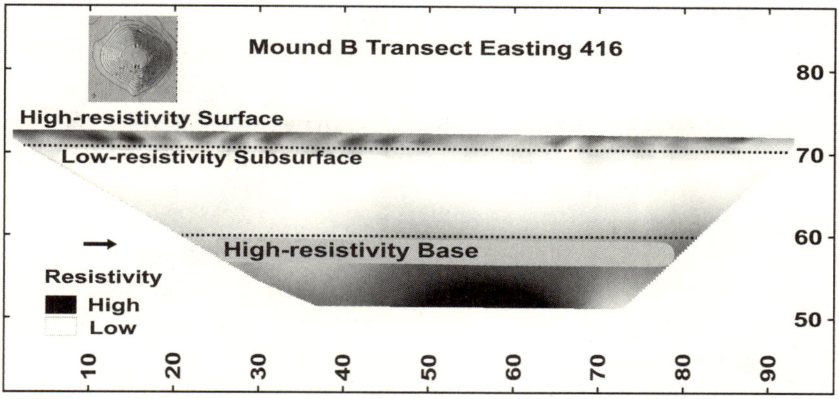


Figure 13.2. Profile recorded off the east side of Mound B, demonstrating a three-layer model of stratigraphy with a conductive layer between two high-resistivity layers.

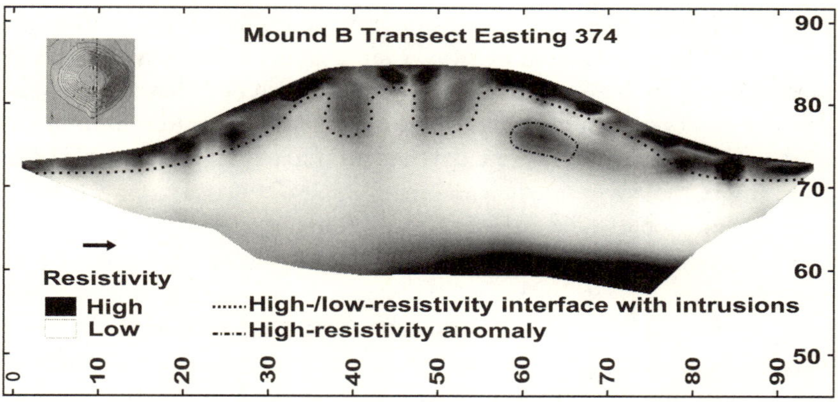


Figure 13.3. This profile recorded across the top of Mound B shows the same three-layer model as Figure 13.2; however, the center layer now also forms the majority of the mound. Two intrusions into this layer around the 40 and 50 m north marks are present in several consecutive transects.

of the mound. At this time, any explanation of these anomalies is speculative; however, trenches full of feasting remnants have been discovered below mound construction phases (Fogelin 2008). Middens and other deposits have demonstrated that feasts occurred at Toltec (Rolingson 2012). It is therefore possible that these trenches could represent evidence of feasting activity prior to the construction of the mound's surface layer.

The upper layer of Mound B remains as highly resistant and dynamic as it is off the mound but becomes much thicker, apparently forming a 3 to 4 m cap on top of the mound. The high resistance and volatility suggest that perhaps several different materials were used in the construction of this phase of the mound, including midden deposits as well as other mound fill. This idea is supported by the data collected from the erosional gully repair. The erosion profile recorded in 1979 revealed 10 deposits, which were grouped into five zones. These zones included layers of midden and mound fill, with silt loam, clay, silty clays, and very fine sandy loam soils. It appears from the resistivity model that the excavation did not extend into the low-resistivity layer of the mound, and the excavation did not discover evidence of the earlier layers.

There is some evidence of the deeper layer in the reports of Edward Palmer, who recorded that he excavated a trench 2.4 m (8 ft) across and 3 m (10 ft) deep. He states in his records of excavations that for the first 0.6 m (2 ft) the soil was "black sticky clay," after which it became "yellow greasy clay," and at 3 m (10 ft) the yellow clay became too hard to work (Jeter 2009). The description of clay throughout is somewhat contrary to the results of the ERT survey and the 1979 profile, once more serving as a reminder that Palmer's descriptions must be used with care. Nevertheless, the description of clay at approximately 3 m, and his decision to stop his excavation at this level, may indicate that he encountered a new material corresponding to the lower-resistivity layer.

A compilation of all the profiles from Mound B in RES3DINV created a fully three-dimensional model that can be manipulated, sliced, and viewed from any direction. The view from above the mound reveals several interesting anomalies (Figure 13.4). The erosion repair on the southern corner is clearly visible as a high-resistivity anomaly. Perhaps as interesting are several linear anomalies of high resistivity at the summit of the mound along the northeastern side and extending toward the center of the mound. It is unknown what these anomalies represent, but it is possible they are related to a garden that was cultivated on the mound summit during the nineteenth and twentieth centuries. Finally, several regions of very high resistivity appear on the northern side of the mound, opposite the plaza. Excavations near mounds on the opposite side from plazas frequently reveal midden deposits associated with feasting (Fogelin 2008; Rolingson 2012). It seems possible that these regions represent such deposits.

No such three-dimensional representation could be generated for Mound A because of the spacing of transects; however, each of the profiles could be analyzed individually. The three-layer model persists in each of these profiles (Figure 13.5). The profile begins at the lakeshore, crossing the north-

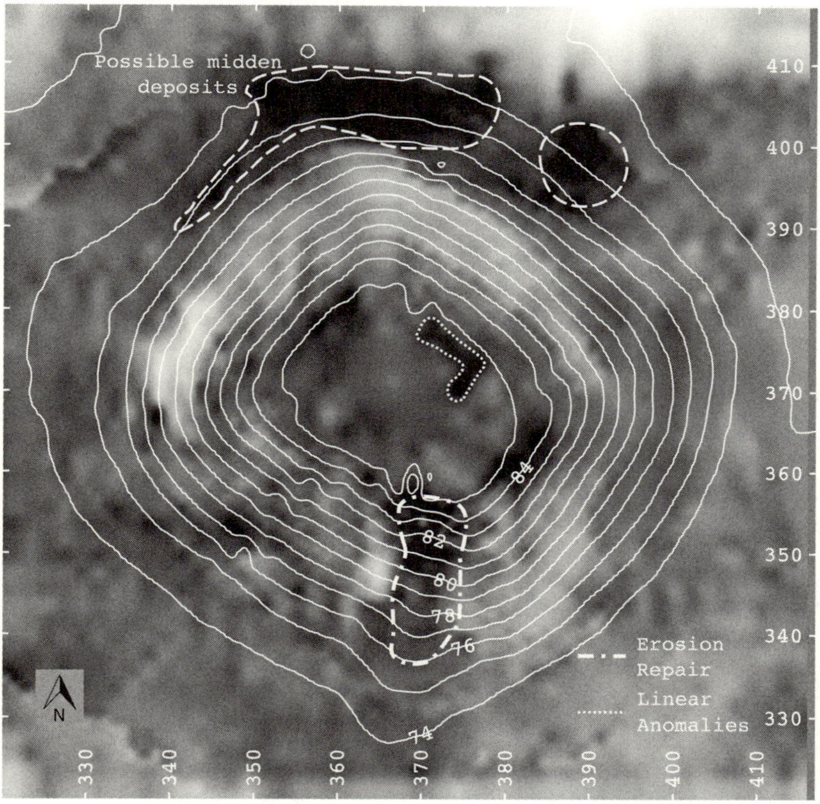


Figure 13.4. Resistivity distribution of the surface of Mound B reveals the erosion cut on the southern corner of the mound and a high-resistivity region on the northern boundary, likely representing midden deposits.

ern edge of the mound and extending out into the plaza. From this it is evident that the high-resistivity base layer indeed represents the water table, beginning at the level of the lake and rapidly dropping away to its standard depth of approximately 5 m below the ground surface. The low-resistivity middle layer is also present in this profile, with the highly dynamic, higher-resistivity layer at the top. The surface layer is also quite thin in the plaza area but is much thicker where the mound is beginning to appear in the profile. Just as the surface of Mound B is covered in 2 to 3 m of high-resistivity material, with the bulk of the structure composed of a low-resistivity material, it appears that Mound A is composed primarily of conductive materials and was also covered in a cap of materials with higher resistivity.

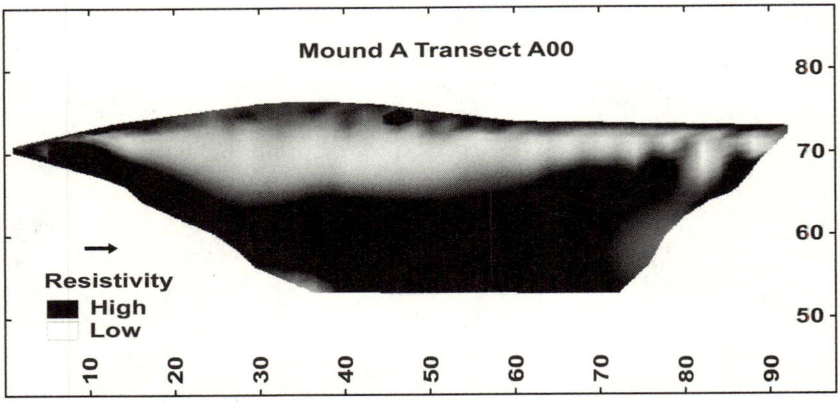


Figure 13.5. This profile taken from the northern edge of Mound A runs perpendicular to the lakeshore. The water table is clearly visible as it descends from the lake.

## Interpretations and Conclusions

The data revealed many anomalies that suggest the presence of archaeological features. Future research will be required if there is ever to be a confident interpretation of the trenches intruding into the conductive layer of Mound B, the rectilinear anomalies at the surface of the mound summit, or the high-resistivity features on the side of Mound B opposite the plaza. However, as was previously proposed, the primary value of remote sensing surveys is no longer just the identification of anomalies. With these data it is possible to propose answers to questions about the construction, morphology, and modification of the mounds, which have potentially dramatic implications for the interpretation of the site and its role in the anthropological understanding of the Terminal Woodland.

For the first time, information about Mounds A and B beneath the high-resistivity cap is available. Neither of the excavations that occurred on Mound B had previously penetrated to this depth, and no commonly used archaeological geophysics techniques were capable of reaching beyond the cap to explore what was beneath. A GPR survey with a 200 MHz antenna was able to penetrate only approximately 3 m into Mound B before attenuating completely. It now seems likely that the rapid attenuation at this depth occurred as the radar waves encountered the highly conductive lower stage of the mound, which simply drained the energy away without providing a return. The ERT reconstruction of the mound's subsurface therefore provides the first indication of the inner structure, hidden by the highly resistant cap.

The presence of a platform mound seems to suggest that the mound may have served a different purpose than conical mounds. The platform may have been intended to hold a structure, such as a "temple" or residence, or to serve as the place for an activity such as a meeting, feast, or speech. In any case, the structure or activity at the top of the mound would almost certainly have been set apart from others occurring on the site. Otherwise the impressive undertaking of building the platform would not have been necessary. Given the limited space at the top of the mound and the presumed importance of the activities that took place there, it also seems reasonable to suggest that the individuals who lived or participated in activities in the space were also in some way set apart from the rest of society, at least for the time they spent there. The platform mound therefore likely suggests an increasing presence of social or political hierarchy, which may or may not be indicated by conical mounds or smaller earthworks.

For these reasons, understanding whether the morphology of Mound B is the result of later modification or whether the mound was originally constructed as a rectangular platform mound is vitally important to beginning to explore the social and political structures of the Terminal Woodland in central Arkansas. One approach to answering this question is to examine the morphology of the newly discovered, earlier stage of the mound. An examination of the profiles (see Figure 13.3) and the three-dimensional reconstruction shows that the earlier stage, though obviously several meters shorter, is also rectangular and appears to have a large flat surface on the top. This would indicate that before the addition of the highly resistant surface layer, the mound was indeed designed as a platform. Furthermore, a radiocarbon sample, collected during the erosional gully repair, produced a one-sigma date of A.D. 806–899, indicating that the cap and therefore the structure beneath it are very likely a product of the Terminal Woodland.

The secondary question of whether Mound A is a severely damaged platform mound then becomes important. It is much more difficult to form a strong conclusion about this without the ability to produce a more accurate three-dimensional model. Nevertheless, it appears that on all sides of Mound A, there is a relatively consistent covering of highly resistant material, similar to Mound B. If destruction had played such a dramatic role in shaping the structure of Mound A and transformed it from a rectangular platform to a conical mound, it may be presumed that in some areas this covering would have been removed. It is reasonable to conclude therefore that Mound A was not reshaped in this way and was in fact constructed much as it appears, in a conical form.

Using ERT, it was possible to answer questions about the morphology and creation of both Mound A and Mound B, which in turn provides one line of evidence that the Terminal Woodland was a period in which society was becoming increasingly hierarchical. Indeed, the Toltec mound complex may so closely resemble the later Mississippian mound complexes because the people who built it were in some ways laying the foundations of power that would arise in the coming centuries. However, ERT did not answer all questions about the mounds. It is still unclear why the tallest mound on the site was not also a platform mound. Likewise, some of the features identified through ERT such as the apparent trenches in the lower portion of Mound B still lack satisfactory explanation. Finally, the necessarily coarse resolution of the survey fails to sufficiently address questions about the evidence of structures or activities that may exist within the mound. Future research in the form of coring, bore-hole magnetometry, and reexcavation of Palmer's trenches may better address some of these questions.

# 5

# Commentary

# 14
# A Decade of Geophysics and Remote Sensing in North American Archaeology

## Practices, Advances, and Trends

*Kenneth L. Kvamme*

It is very pleasing to see a new book on the state of archaeological geophysics in American archaeology. The chapters herein offer an opportunity for commentary on how it is practiced today by a new generation of researchers. The chapters reveal the state of the art in field methods, instrumentation, software, data-processing techniques, forms of presentation, and even theoretical orientations. In addition, they testify to the incredible nature and variety of archaeological discoveries made possible by archaeogeophysics from across the land. All are significant.

A decade ago the first collective statement of North American practice in archaeogeophysics was offered in the volume edited by Jay Johnson (2006), *Remote Sensing in Archaeology: An Explicitly North American Perspective*. In essence, it was a how-to text describing the methods and theories of near-surface geophysical prospection. At that time this was necessary because geophysical methods were nearly off the radar screen of most American archaeologists. Few knew about it or employed it, and programs of training in the topic were rare within anthropology departments.

The present volume offers a major departure. With a few exceptions, it is little concerned with instruction or the explanation of geophysical principles. That has already been done. Its primary concern lies in applied archaeogeophysics as a set of tools for the discovery and mapping of buried archaeological features and for addressing anthropological and even historical questions.

In this vein, this cross section of geophysical results from the United States illustrates aspects of prehistoric and historic archaeology that were heretofore not considered and perhaps not even realized by archaeologists. Examples include the layouts of entire hidden villages and hamlets, precise counts of houses or even food storage pits, and even the locations of long-buried and long-forgotten Civil War entrenchments. In this regard, this volume represents the first collection focused on geophysical *results* from across the United States, as opposed to geophysical *methods* and the instructional concerns that have characterized previous compilations. Indeed, the results-based orientation of this volume is reflected in its organization around four anthropological themes.

At the same time, the instrumentation, methods, theoretical orientations, and justifications for doing geophysics in this work provide us with a "snapshot" of the state of practice in North America in the second decade of the twenty-first century. From a historical point of view, this document will represent our practices to the future, and it also provides us with a referent for comparisons against past approaches.

These perspectives raise a multitude of questions. For example, what kinds of instrumentation and field methods tend to be employed today? What are the principal concerns in data acquisition? What kinds of things do projects hope to learn? How are surveys justified? In addition, my research interests forced me to focus on any developments in multisensor surveys and their integration, uses of GIS, how geophysical results are interpreted, and whether geophysical studies might stand alone as primary data to address anthropological research questions (Kvamme 2007a; 2007b; 2008b).

## Geographic Focus

Perhaps the simplest place to start is with geography. Specifically, where is archaeogeophysical work in the United States being conducted? Most of the results in this volume come from the Southeast, broadly speaking, and this is probably not a coincidence. A recent special issue of the journal *Archaeological Prospection* (volume 21, issue 1, 2014) entitled "Exploring the Space Between: Remote Sensing, Monuments, Households, and the Broader Landscape in the American Southeast" testifies to the preponderance of work in this region. More geophysical surveys are likely conducted there in part because of the nature of the area's impressive archaeology, with large settlements and mound complexes, its largely unknown state (from a geophysical standpoint), and its growing population density and associated development. Of the 17 sites or study areas discussed in this volume, 10 are in the Southeast alone, and perhaps 11 if we throw in the nearby Lower Illinois River Valley. The remainder of the country is represented by a sprinkling of studies from the Southwest, Great Basin, Great Plains, and northern Midwest.

## Instrumentation

For the most part, geophysical survey methods in use now are similar to those that have been employed for the past couple of decades in North

America, although there are a number of important departures and perhaps some shifts in popularity. Motor-driven multiarray carts for fast surveys of large areas, each containing a suite of fluxgate gradiometers or perhaps ground-penetrating radar (GPR) antennas, are only starting to be seen in American practice (e.g., Davis et al. 2015). They have been more commonly employed by larger European institutions and reflect the state of the art in instrumentation (Gaffney et al. 2012).

It is perhaps unsurprising that magnetic gradiometry surveys dominate this collection, with use in 9 of the 12 chapters that report ground-based survey results. This is partly due to their speed and rapid coverage of large areas in comparison with other geophysical survey methods, yet these characteristics would be meaningless if magnetic gradiometry did not consistently deliver useful results. Indeed, anthropogenic alterations within archaeological sites tend to produce magnetic variations that are plentiful and highly interpretable, making this property "nature's gift to archaeology" (Kvamme 2006b).

GPR has increased in prevalence, with use in six of these chapters, and for many excellent reasons. The past decade has seen advances in survey speed through the near-universal adoption of wheeled carts. Improvements in software have made consideration of large data sets possible, and recent guides facilitate interpretation of the data (Conyers 2013, 2014).

Lateral resistivity surveys seem to have lost their popularity, with applications in only four chapters. Despite the generally good results this method offers and the introduction of speedier cart systems more than a decade ago (Somers 2006), few have adopted it, and the twin-probe array remains in use despite its slow rate of data capture. At the same time, one of the surprises of this volume is the two chapters that utilize electrical resistivity tomography (ERT), a method nearly unused in American archaeogeophysics until very recently (it was barely mentioned in J. Johnson 2006).

Electromagnetic induction (EMI) instruments were touted by Clay (2006) and others as "upcoming" a decade or more ago because quadrature phase ground conductivity data (the inverse of resistivity) could be acquired much more rapidly over large areas than could corresponding resistivity data through probe-contact surveys. At the same time, in-phase data from an EMI instrument, which quantifies near-surface magnetic susceptibility (MS), were little emphasized in the past except by Dalan (2006). In the present volume only a single chapter makes use of ground conductivity data in a rather small (18 x 18 m) study (Hammerstedt et al. in chapter 1), and the results offer few insights over those of the other survey methods employed. In-phase MS data are obtained in two chapters, however, with clear results. When combined with three MS data sets acquired with a single-coil device (Lynch and Becker, chapter 9), the results prove to be another of this volume's highlights and represent a major sea change in contemporary practice (more on this later).

The application of airborne LiDAR technology by Thompson (chapter 8) for topographic mapping of cultural features expressed on the surface is relatively new in American archaeology (see, however, J. Harmon et al. 2006). This chapter represents a departure from the others, all of which employ ground-based methods. LiDAR was becoming available only a decade ago and the high-resolution data sets (of about 1 m) necessary for most archaeological work remain rare and expensive to acquire in much of the country, although usage is growing rapidly.

## Multimethod Surveys

Multimethod surveys are common in this volume, as they also were a decade ago. Yet it is surprising that as many as six of the chapters are based on a *single* geophysical method. Admittedly, these studies are generally linked with such ancillary data as digital elevation models, aerial imagery, or mappings of surface artifacts, helping to make them more than one-dimensional (several studies also reported that other geophysical results were not included in order to streamline their presentations). Of the remaining seven chapters of ground-based surveys, five employ two or three methods, and two utilize as many as four.

It is of some interest *why* multiple geophysical methods are employed. Hammerstedt et al. (chapter 1) explain that "results suggest that our use of multiple technologies was crucial to our interpretation of the data." While this is certainly true in many instances, it may not always be the case. Often there is a real purpose for conducting a survey with an additional geophysical method—for example, to better define subtly indicated house floors after a gradiometry survey by their increased electrical resistivity or the likelihood of robust microwave reflections seen in GPR (e.g., see Hammerstedt et al., chapter 1; King et al., chapter 11). On the other hand, in some cases it appears that an instrument and the necessary labor may have been available and so the survey was run for that reason alone, with results that contribute little to the understanding of the subsurface. Often, a second survey of an area with a different geophysical method offers only validation of an anomaly already indicated by a prior survey. This circumstance is sometimes cited as a justification for conducting additional surveys, but if the initial survey was conducted properly, do we need confirmation that an anomaly is really there? The question of how much more information additional surveys deliver over an initial survey cannot be easily answered at this time and will probably require a great amount of experience in various natural and archaeological settings.

Although multiple studies employ several geophysical methods over the same piece of ground, no attempts at digital data fusion were made that might enhance subsurface portrayals by combining the information into a single composite image through mathematical, statistical, graphical, or other means (see Kvamme 2007b). Rather, side-by-side mappings of results or image overlays are employed to depict multisensor findings, commonly through the use of GIS technology.

## GIS and Geophysics

A decade or so ago the use of GIS for the management and integration of geophysical data was only in its incipient stages (e.g., Neubauer 2004). In the present volume, GIS is often used for mapmaking purposes, but much more is made of data integration between geophysical and other types of information, such as surface artifact densities, digital elevation data, or aerial imagery. The evidence is convincing. At the Spiro Mound Group in eastern Oklahoma, Hammerstedt et al. (chapter 1) explain an apparent texture difference in magnetic gradiometry data by noting its association with a soil change. McKinnon and Haley (chapter 3) demonstrate interpretational insights gained by combining surface artifact distributions with geophysical results. Wiewel (chapter 6) does much the same with digital elevation data by showing its critical relationship to the clarification of magnetic gradiometry findings. Sturm's (chapter 5) analysis of subtle evidence for agricultural fields at Chaco Canyon in New Mexico demanded precise coregistration of findings from multiple geophysical, topographic, and aerial data sets that "could not have been achieved" without GIS.

## Large-Area Surveys

This volume demonstrates that much larger ground-based archaeogeophysical surveys are now often contemplated. Patch et al. (chapter 2) assert that "large-scale geophysical surveys represent a new and productive approach," and a recurrent justification is the need to visualize sites at the landscape level. Several of the presented surveys are enormous by American standards, with 30 ha reported at Spiro Mounds, Oklahoma (Hammerstedt et al., chapter 1), and several other surveys achieving respectable sizes between 5 and 12 ha. All were performed with magnetic gradiometry, without the benefit of motorized multiarray carts that have generated kilometer-sized surveys in Europe (Gaffney et al. 2012). Obviously, survey speed is a central concern and, indeed, it is the increased speed and memory capacities of modern instrumentation that make large surveys possible. All of the large magnetic gradiometry surveys made use of the fast Bartington Grad601, an instrument that was making inroads in the United States only a dozen years ago. Its twin fluxgate gradiometer tubes permit simultaneous data acquisition in two transects for every one walked by the operator, which instantly doubles survey speeds (and therefore areas).

It is not surprising that GPR surveys have also grown much larger thanks to the introduction of wheeled cart systems and improved software that can handle the large data volumes generated. Patch et al. (chapter 2) show GPR slice maps that cover nearly 3 ha and reveal a host of patterned anomalies in a Mississippian mound center in Tennessee. The speed of these surveys does lag behind that of magnetic gradiometry owing to the need to push a cart, and this is especially true over rough ground. Moreover, despite software advances, data processing remains cumbersome, in part because the data are three dimensional. At a minimum, radargrams showing vertical profiles as well as horizontal plan maps or "time slices" must be generated and examined. The five other chapters that utilize GPR perform it in small patches that measure much less than a hectare each. In some cases, this is due to predefined small survey areas or, at Chaco Culture National Historical Park, to a wish to minimize damage to fragile vegetation (Sturm, chapter 5). Yet the tendency is evident in this volume to employ GPR for focused work in predefined areas of interest based on such other information as the presence of a mound (King et al., chapter 11; Bigman and Seinfeld, chapter 12) or interesting anomalies indicated in a prior magnetic gradiometry survey (Patch et al., chapter 2; Henry et al. chapter 10).

One form of large-area survey that will surprise experienced readers is magnetic susceptibility, which was carried out with the Bartington MS2-D, a single-coil instrument known for its painfully slow rate of data capture and limited depth penetration (less than 10 cm; Dalan 2006). Lynch and Becker's (chapter 9) survey of historical Fort Bridger in Wyoming covers perhaps 8–9 ha, albeit at a low sampling density of a single measurement every 5 m. These results along with those in two additional, but much smaller, case studies at other sites are excellent and represent something new in one of the more important contributions to this volume.

## What's New—Methods and Approaches

Some of the most distinctive characteristics of this volume lie in several unusual archaeogeophysical methods or approaches that are employed, at least from an Americanist standpoint. Already discussed is the integration of geophysical information with such other data sets as microtopography and aerial imagery via GIS. Also included are several geophysical methods that saw rare use only a decade ago and an important one that has seen little use in American applications.

## Magnetic Susceptibility Surveys

As alluded to in the previous section, the large-area MS surveys of Lynch and Becker (chapter 9) represent something novel, and not just to American practice. MS surveys are certainly not new in this country. A decade or more ago Dalan (2006) and others offered a number of exciting and even pioneering results. What is innovative here is (1) its application with uniform sampling to unusually large areas—even entire settlements (rarely performed, but see Gaffney et al. 2005), (2) an interpretational perspective that equates a mapping of variations in MS with changing intensities of human activity across a site (with high MS pointing to greater activity), and (3) a perspective change that sees its use as a primary archaeogeophysical tool for wide-area mappings. Lynch and Becker not only define activity areas within historical Fort Bridger (8–9 ha); they also show likely professional and domestic activity areas related to the life of a lighthouse keeper in 1847 in northern Michigan (in a survey of 0.1 ha). In another case study they also explore the MS characteristics of a historical tar kiln in North Carolina (0.035 ha), where we are also treated to one of the first applications of area-wide and down-hole mapping of variations in magnetic viscosity.

In all cases, the slow, single-coil Bartington MS2-D is employed, which is a taxing choice for large-area surveys. The MS2-D requires perhaps 10 seconds for both the "air" reading necessary for calibration and the actual ground measurement. In addition, because this shallow-depth instrument (with a depth range of less than 10 cm) is so impacted by distance variations between the coil and the ground caused by intervening vegetation, Lynch and Becker (chapter 9) took several readings at each station and then averaged them, requiring even more time.

Twin-coil devices such as the Geonics EM-38 series permit rapid measurement of MS in the in-phase component at a rate of several per second and at greater depths, perhaps 50 cm (Dalan 2006, 2008). Yet until recently, they have been little used for large-area MS mapping. Two such studies are offered in this volume, not to quantify activity or to cover large areas, but as a means of subsurface discovery and validation of findings yielded by other surveys. Hammerstedt et al. (chapter 1) show a 0.16 ha area mapped by a twin-coil device to corroborate subtle "structure-like" anomalies indicated by their magnetic gradiometry survey at Spiro Mounds, Oklahoma. Similarly, Henry et al. (chapter 10) employ an EM-38A to map MS over a Civil War rifle pit in a 0.08 ha survey that gives a clearer result than their magnetic gradiometry findings.

## ERT surveys

This volume presents some of the first published applications of ERT surveys in American archaeology, and their great potential is clearly indicated. They have been quite popular in European archaeology for a number of years, with often stunning results (e.g., Papadopoulos et al. 2014; Papadopoulos and Tsourlos 2007). Unlike the geophysical methods of other chapters, which are confined to relatively shallow depths of a few meters, ERT permits deep-earth soundings of tens of meters in the form of pseudosections and even three-dimensional (3D) models of the subsurface (Schmidt 2013). Such operations are necessary to explore archaeological mounds in depth, a large focus in this volume and important to archaeologists in the eastern half of the United States. King et al. (chapter 11) offer comprehensive insights into mound details in the Lower Illinois River Valley through ERT profiles. They document features consistent with expected internal ramps and tombs in a shallow mound that GPR results and magnetic gradiometry findings seem to support. Zimmer-Dauphinee (chapter 13) goes a step further with three-dimensional modeling of the interior of Toltec Mound B, a rectangular platform mound that is one of the largest in Arkansas. ERT sections address an important question about whether Mound B was originally constructed as a platform mound. It is revealed that an earlier stage of construction, several meters smaller, was also shaped as a flat-topped platform.

## LiDAR (Light Detection and Ranging)

Aerial LiDAR surveys have come to the attention of American archaeology only in the last decade, and their use tends to be rare owing to their expense and the limited areas of the United States that have been subjected to high-resolution and publicly available surveys. The single chapter on LiDAR by Thompson (chapter 8) may seem incongruous with the remainder of this volume, but its presence is important because this method is only now beginning to make an impact in the archaeology of the Americas (see J. Harmon et al. 2006 for an excellent introduction). Revolutionary results akin to those achieved throughout European archaeology are already occurring, such as the recent mappings of the ancient Maya civilization in the jungles of Central America (Chase et al. 2011).

In Thompson's presentation we are informed about unusually large settlements in southern Florida. Despite what appear to be low-resolution data (as indicated by the illustrations) compared to those of most archaeological applications of LiDAR, general structural details are visible in the kilometer-sized scales of some of the settlements. Some of the first images of the gigantic shell-bearing sites (measuring from 23 to 51 ha) are offered that appear to contain prepared mounds. The scale of these sites is stunning and arguments for their organizational layout are given based on characteristics of symmetry, axiality, and orthogonality. It is unstated, but one wonders whether these sites occur in wetlands and in highly vegetated settings as some of the discussion seems to imply. Even if this is not the case, their sheer size makes learning about them through other geophysical methods a daunting proposition indeed. Whatever the case, LiDAR seems best suited for gaining initial understanding.

## Reasons for Geophysical Surveys

Several reasons are indicated for conducting archaeogeophysical surveys. Foremost among them is site identification, the discovery of buried archaeological sites and features. Locating buried features was the original purpose of geophysical surveys in archaeology, often for possible excavation (see Clark 1996). Today, they are used more to locate and map subsurface features. All of the chapters dealing with ground-based prospecting follow this pursuit in one way or another.

A related focus that follows from the first is the identification of intrasite structure and community plans or layouts by the extensive subsurface mapping of features. This pursuit has occurred only in the last couple of decades as surveys have become large enough (thanks to faster instrumentation and larger memory capacities) to yield results that can encompass large villages. Most chapters focus in this area too, except those that concentrate solely on specific features within sites, such as mounds.

Visualization of sites at the landscape level, an often-cited reason for conducting large geophysical surveys, has come to fruition only in recent years with improved instrumentation and implies extremely large survey areas that typically go beyond the extent of a single site. This permits the site to be viewed in a landscape context and its relationships with other sites to be examined (e.g., see Gaffney et al. 2012). In the present volume, although several large surveys are conducted, only the largest barely qualify, because most studies focus on single sites. In McKinnon's description of the Battle Mound site, his wide-area geophysical results from some 12 ha do indicate discrete and widely dispersed activity areas that probably point to separate occupational areas within this large landscape. The survey by Hammerstedt et al. (chapter 1) is even large at 30 ha, but it occurs within a single site, the ceremonial complex of Spiro Mounds. Nevertheless, given the scale of this important ancient center and the landscape variation evident in the topographic mapping, this study certainly qualifies as a landscape study. Airborne LiDAR surveys offer the classic example of a landscape archaeology perspective because data are rapidly acquired over square kilometers and highly detailed indications of cultural resource sites are included (Chase et al. 2011). However, Thompson's (chapter 8) application of LiDAR, like many of the volume's other studies, is site focused, offering little about landscape-level patterning.

Another purpose for archaeogeophysical surveys is apparent from this book. Patch et al. (chapter 2) suggest that it is important to *reinvestigate* sites of former large archaeological projects, such as those conducted by the Works Progress Administration (WPA) in the 1930s. Those projects performed large excavations that generated huge collections, but their field notes, plans, and maps do not meet current professional standards and their interpretations remain incomplete. Geophysical surveys of these sites, when combined with the WPA collections and notes, can greatly augment our understanding of these sites and the prehistoric cultures they represent and help confront issues related to their better management.

Lynch and Becker (chapter 9) and Patch et al. (chapter 2) also emphasize that geophysical survey data can aid eligibility recommendations and future nominations for the National Register of Historic Places. In addition, the knowledge of the subsurface they impart can produce recommendations for best management practices. Both are important justifications for archaeogeophysics and parallel the framework established by Lockhart and Green (2006) in their pioneering summary of the roles of remote sensing in cultural resource management.

## Interpretations of Anomalies

Most of the studies in this volume are site or feature focused (e.g., mounds) and begin with a detailed discussion of the sites, their cultural and excavation histories, and other background information. Geophysical surveys are generally brought in to augment understanding of them. Overall, a primary focus lies in feature discovery and in mapping the subsurface. A major goal is the production of high-quality imagery that can aid in developing interpretations of anomalies that represent likely cultural features. The methodology of interpretation is based largely on pattern association through the shapes, relative sizes, and spatial arrangements of the anomalies. For example, those that are small and circular might indicate hearths or storage pits, while square or circular anomalies of a larger size might point to houses, and large linear anomalies might represent fortifications, trails, walls, and the like, depending on where they are placed. A large part of this inferential guidance comes from plans, sections, and photographs gained from the corpus of traditional excavation materials acquired over the past century. The end product is some sort of classification of anomalies into a series of projected archaeological feature types.

A rarer and more deductive form of interpretational reasoning is also seen in this volume and is best exemplified by Lynch and Becker (chapter 9) in their discussion, analysis, and interpretation of the varied magnetic properties of a historical tar kiln at Clear Run, North Carolina. They dissect tar-making practices, kiln form, and the effects of reducing and oxidizing environments on magnetic enhancements of the soil and employ magnetic gradiometry, susceptibility, and viscosity surveys (including down-hole methods) to measure them and derive a series of new insights that explain anomaly forms seen in the data.

Because archaeogeophysical surveys are now large and frequently focus on settlement sites, thousands of classified anomalies are common and new discoveries about the nature and content of the sites are a general rule. In fact, with the large exposures currently possible through archaeogeophysics, we are now in a new age of archaeological discovery the likes of which we have not seen since the massive WPA excavations of the 1930s. It is inevitable that occasionally something new is found that has not been seen before in either geophysical or excavation results. Such examples are relatively plentiful in this volume. The unexpected evidence of a "fence" surrounding a pithouse cluster at the Dillard Site (Diederichs et al., chapter 7), the presence of circular houses at Huff Village (Wiewel, chapter 6), the rectangular geometry in a Middle Woodland mound structure (King et al., chapter 11), and the unexpected buildings near Craig Mound at Spiro (Hammerstedt, et al., chapter 1) are examples that come to mind. Despite these positive outcomes, all emphasize how much we have yet to learn about the archaeology of North America, and geophysical surveys represent the tools that will make it possible by their lower cost (compared with excavation), their large areas of exposure, and their noninvasive character.

The chapters in this book demonstrate an increased confidence in the interpretation of geophysical anomalies—what they might represent archaeologically. In chapter 2, for example, Patch et al. make very specific archaeological assignments to *thousands* of anomalies across two large Mississippian village and mound centers in Tennessee and Alabama. Interpreted pits, middens, structures, posts, mound features, gate features, and other cultural constructions are identified based largely on anomaly forms. McKinnon and Haley (chapter 3) do much the same in respective Caddo and Mississippian mound centers, although they are much more conservative in the archaeological categories they are willing to claim, focusing only on mounds, structures, cemeteries, farmstead compounds, and borrow pits. Wiewel (chapter 6) confidently counts nearly 2,000 small circular anomalies in village-wide magnetic gradiometry as food storage pits to make predictions about agricultural productivity and, ultimately, census. In his case, previous archaeological testing and coring were correlated with the geophysical results to provide more confident interpretations. Bigman and Seinfeld's (chapter 12) GPR study of Late Woodland mounds in Florida asserts that mounds are known to contain human burials and that consequently, anomalies in mounds may point to burials. While this logic cannot be faulted and provides a conservative starting point for interpretations, they admit that testing through excavation is unlikely given laws protecting human burials. They make a stab at identifying certain hyperbolic reflections as burials, with convincing evidence in several cases, but one wonders how many hyperbolas may point to graves given the internal complexities of mound construction, fills, tree roots, and other disturbances.

Some of this general confidence may arise from the increased corpus of work in American archaeogeophysics over the past two decades that has established a groundwork of reference material facilitating interpretations of anomalies (e.g., as expressed in J. Johnson 2006 and elsewhere). Yet in most cases such interpretations remain untested.

## Benefits of Noninvasive Investigations

Another recurrent theme for doing archaeogeophysical surveys, seen throughout the volume, is their noninvasive nature, a characteristic that has long been recognized (e.g., Heimmer and De Vore 1995:1). What seems different now is this feature's core importance to the conduct of archaeology in a variety of contexts. Four of the chapters (1, 4, 5, and 13) report work in national or state parks where preservation and protection of cultural resources is a primary mission and a growing demand exists for noninvasive or minimally invasive archaeology. At the same time, excavations are increasingly limited by funding, national and state cultural resource regulations, and a discipline-wide increase in sensitivity to Native peoples. Geophysical surveys are therefore seen as an alternative to excavations because maps of subsurface features and even settlement layouts may still be generated with minimal disturbance to the ground. Mink (chapter 4) observes that with a limited ability to excavate, one of the biggest challenges of geophysical work in a national park is confirming the source of an anomaly. He argues for conducting geophysical surveys on all sites before excavations that may be required by federal compliance processes. This information could then be used for the development of an "anomaly signature" database for application to sites *not* excavated as a way to improve interpretations. Such an idea should be of national importance both inside and outside parks.

Most of this volume reports research outside national and state parks, of course, and even in these contexts similar issues arise. Patch et al. (chapter 2) applaud the use of geophysical surveys for site discovery in part to minimize the volume of expensive curation collections resulting from traditional excavations. Bigman and Seinfeld (chapter 12) and King et al. (chapter 11) observe that laws regulating the disturbance of human burials and changes in archaeological ethics have greatly limited mound archaeology. With direct inspection of mound interiors through excavation rarely possible, their investigation *must* utilize geophysical methods. From these arguments, the growing importance of geophysics as a replacement for excavation should be realized in cases where legal, ethical, or cost considerations prohibit digging.

## Anthropological Interpretations and Questions

In the twentieth century the development of archaeology as a discipline witnessed "Descriptive" and "Classificatory-Historical" periods, which were followed by an era, beginning about 1960, that remains characterized by "Explanation" (Willey and Sabloff 1993). The last includes a greater anthropological viewpoint as well as processual concerns for problem-oriented research in a scientific framework and a more recent postprocessual reaction. The archaeogeophysical research presented in this volume mimics much the same framework. For instance, most work remains focused on *describing* and *classifying* the nature and location of subsurface archaeological features, and rightly so. This was the initial core reason for the use of archaeogeophysics in archaeology. A number of chapters see geophysical surveys in a different light, however, where the primary emphasis lies in using the data to test hypotheses and to address anthropological questions, often from a processual perspective. One chapter even offers a postprocessual viewpoint where the motivation rests in how contemporary "agents" (i.e., people) connect with a place and its past and in how archaeogeophysics can address those needs.

A characteristic that stands out in this volume is a movement away from the notion that geophysical methods are only tools for subsurface feature discovery. There is much discussion about using geophysical results for anthropological interpretations and, indeed, the volume is organized around four anthropological themes. At the simplest, an anthropological interpretation seems to mean using geophysics to identify *anthropogenic features*. In other words, because geophysical surveys can reveal such constructions as pits, houses, ditches, plazas, and the like, and these features are built by people, the results are therefore "anthropological," a usage indicated in multiple chapters.

More significant is the idea of addressing anthropological *questions*. Wiewel (chapter 6) provides the best exemplar of this perspective, with specific questions about the number of storage pits, storage capacities, levels of agricultural production, per capita storage volumes, and population size at Huff Village in North Dakota. McKinnon and Haley (chapter 3) do not pose specific questions, but they *describe* interpreted features in two Mississippian settlements from an anthropological standpoint by focusing on the distribution of architecture and uses of community spaces. The several mound researchers pose similar, if less demanding, questions. Bigman and Seinfeld (chapter 12) ask whether smaller mounds in Florida were constructed in one or several episodes and whether any contain burials or architecture. Zimmer-Dauphinee (chapter 13) wants to know whether an earlier mound had a "platform" shape within Mound B at Toltec Mounds in Arkansas.

These case studies illustrate that geophysical results can be used directly, as the primary information in a project, to address research questions of an anthropological nature. Even better, geophysical surveys permit questions that are impossible to confront through excavation. The large areas or volumes of the subsurface that can be imaged by geophysics enable research questions and analyses related not to the confines of a house or two, but to *all* houses within a village, and this represents a revolutionary sea change in the conduct of settlement or landscape archaeology.

Although arguably anthropological, many of the foregoing questions and usages reflect relatively easy-to-identify characteristics of sites and their features. For the most part they address less complicated questions such as numbers of storage pits and their capacities, and evidence for earlier mound forms and their shapes. This volume demonstrates that we have become proficient at obtaining such evidence. But what about higher-level anthropological questions that relate to theory building, understanding of sociopolitical complexity, evolution and growth of village structures, cultural change, and the like? Patch et al. (chapter 2) suggest possibilities in this pursuit with the observation that "site size and feature density can be used to infer the intensity of sociopolitical complexity." Perhaps examination of feature counts and densities, the multiple components of house forms, the presence or absence of specific architecture types, organizational layouts in mound or plaza distribution, the nature and types of defensive systems, and the like could ultimately yield indices relevant to determinations of levels of social complexity. In the end, however, pursuing higher anthropological questions may rest on the relatively simple identification of features and forms that geophysical surveys currently offer.

In a departure from the more conventional approaches of this volume, one of the first chapters in American archaeogeophysics written in a postmodern or postprocessual perspective is offered, and it presents something very new. Henry et al. (chapter 10) describe the idea of "placemaking," the ability of geophysical surveys to reveal past places of cultural significance that can give the public a greater connection to its past. As they put it, "reversing placelessness [the loss of place identity] and providing local communities a mnemonic by which they can attach themselves to local landscapes and engage in the practice of placemaking is one role archaeological geophysics is poised to serve." In a case study in Kentucky from the American Civil War, it is not surprising that after nearly 150 years there is little social memory of the battle, as the actors, and indeed several generations of their descendants, are long dead and a replacement community of new immigrants has moved in. Nevertheless, this chapter represents an important new angle for the justification of geophysical surveys, and one that is sure to grow in importance.

## Projections for the Future

What will the future bring to our still-youthful practice of American archaeogeophysics? One is always hesitant to step into such unfamiliar territory, if only because many projections look so wrong only a decade or two down the road. A safer course of action is to mix expectations for the future with hopes.

Perhaps Mink's (chapter 4) call for a geophysical "signature" database to enable the linkage of known archaeological feature types with specific anomaly forms revealed by our surveys should come to fruition. Such an undertaking is likely only within a large institution, such as the National Park Service. Such databases are certainly needed, especially with the increasing tendency seen in this volume to substitute excavation with geophysical study owing to legal, ethical, preservation, and cost issues.

Undoubtedly, the use of instruments in the field will continue to be made easier as menu-driven systems, automatic parameter settings, more intelligent defaults, and the like advance. With minimal instruction in how to move an instrument over the ground (now also observable on YouTube), reasonably intelligent individuals are able to perform their own surveys with little formal training or mentorship. Indeed, several courses and field schools across the country now provide such training. In the past decade, data-processing software has markedly improved and will continue to do so. Downloading data from field instruments is already a snap, and with increasingly standard recipes for data processing, reasonably good images of magnetic gradiometry results to once-complicated GPR time slices may be produced. A major goal of most field applications, as this volume illustrates, is only a map that may point to potential subsurface archaeological features. Given these circumstances, geophysical surveys that only a decade ago were uniformly conducted by specialists will likely evolve into a much more common operation that will be little more demanding than approaching a field computer and entering data in a GIS.

Large-array carts for acquiring magnetic gradiometry or GPR data over large areas (Gaffney et al. 2012) are certainly warranted given the sizes of some of the surveys seen in this volume. Yet given their cost, such a development seems unlikely unless government agencies that manage large properties (such as the National Park Service and Department of Defense) decide to employ archaeogeophysics at very large scales.

Multimethod surveys are sure to increase in number, but with improved consideration of which methods work best together under various conditions as opposed to the haphazard trials and shortcomings occasionally seen today. GIS-based data integrations will also increase, especially into the domain of satellites as newer platforms with greater spatial and spectral resolution are released, such as WorldView 3 with its 16 bands and spatial resolutions as high as 0.31 m (Kruse and Perry 2013). Digital data fusions for enhancements of information content are sure to follow by combining multisensor information, perhaps from ground, air, and space through mathematical, statistical, or graphical means.

Software evolution also deserves comment in that something akin to a major leap in complexity and sophistication has occurred in recent years, particularly in the arenas of GIS and statistical computing. This software is a boon to productivity and to confronting many-stepped and complex operations, and this is particularly true for those researchers who have excellent training in these respective fields. Yet modern software is also a set of point and click–directed "black boxes" where a single click can launch a suite of complex operations that, increasingly, the clicker little understands. While there is some possibility of deducing what is going on by witnessing cause and effect—for example, the state of the data before and after a click—such systems offer potential dangers. It can be the case that an end product of such automated computation is replete with processing artifacts resulting from less than ideal choices. As we move ever further toward greater automation in archaeogeophysics—perhaps combined with less training—it can only be hoped that some wariness will remain as well as a core of practitioners who can recall and reconstruct basic principles and processing fundamentals. The studies reported in this volume give hope to such a future.

# References


Abbott, James T., R. Blake Roxlau, and William B. Butler

1997 *Kirtland Air Force Base: Geophysical Studies.* TRC Mariah Associates, Inc., Albuquerque, New Mexico. Submitted to Kirtland Air Force Base and National Park Service, Denver, Colorado.

Abernathy, Edward J., Karen M. Olszewski, and Rodney Peters

1983 *Soil Survey of LeFlore County, Oklahoma.* USDA Soil Conservation Service, Washington, DC.

Ahler, Stanley A.

1993 Plains Village Cultural Taxonomy for the Upper Knife-Heart Region. In *The Phase I Archeological Research Program for the Knife River Indian Villages National Historic Site, Part IV: Interpretation of the Archeological Record*, edited by Thomas D. Thiessen, pp. 57–108. Occasional Studies in Anthropology No. 27. Midwest Archeological Center, National Park Service, Lincoln, Nebraska.

2000 Radiocarbon Analysis. In *New Geophysical and Archaeological Investigations at Huff Village State Historic Site (32MO11), Morton County, North Dakota*, edited by Stanley A. Ahler and Kenneth L. Kvamme, pp. 63–66. Research Contribution No. 26. Paleo-Cultural Research Group, Flagstaff, Arizona. Submitted to State Historical Society of North Dakota, Bismarck.

Ahlman, Todd M., Susan R. Frankenberg, and Nicholas P. Hermann

2000 *Archaeological Reconnaissance Survey of Tennessee Valley Authority Lands on the Watts Bar Reservoir.* Tennessee Valley Authority, University of Tennessee Department of Anthropology, Knoxville.

Anderson, Kirk C., and Ted Neff

2011 The Influence of Paleofloods on Archaeological Settlement Patterns during A.D. 1050–1170 along the Colorado River in the Grand Canyon, Arizona, USA. *Catena* 85:168–186.

Anschuetz, Kurt F.

2006 Tewa Fields, Tewa Traditions. In *Canyon Gardens: The Ancient Pueblo Landscapes of the American Southwest*, edited by V. B. Price and Baker H. Morrow, pp. 57–74. University of New Mexico Press, Albuquerque.

Anschuetz, Kurt F., Richard H. Wilshusen, and Cherie L. Scheick

2001 An Archaeology of Landscapes: Perspectives and Directions. *Journal of Archaeological Research* 9:157–211.

Arrhenius, Gustaf, and Enrico Bonatti
1965 The Mesa Verde Loess. In *Contributions of the Wetherill Mesa Archaeological Project*, assembled by H. Douglas Osborne, pp. 92–100. Memoirs No. 19. Society for American Archaeology, Washington, DC.
Aspinall, Arnold, Chris Gaffney, and Armin Schmidt
2008 *Magnetometry for Archaeologists*. AltaMira Press, Lanham, Maryland.
Atalan Çayirezmez, Nurdan, Pinar Ertepinar Kaymakçi, and Geoffrey D. Summers
2008 Remote Sensing Methods at Kerkenes: Combining Geophysical and Other Methods. *Journal of the Earth Sciences Application and Research Centre of Hacettepe University* 29:87–100.
Bailey, Garrick
1995 *The Osage and the Invisible World: From the Works of Francis La Flesche*. University of Oklahoma Press, Norman.
Baker, Frank C., James B. Griffin, Richard G. Morgan, Georg K. Neumann, and J. L. B. Taylor
1941 Contributions to the Archaeology of the Illinois River Valley. *Transactions of the American Philosophical Society* 32, Part 1.
Baker, Henry A.
2005 *Archaeological Testing at the Letchworth Mounds Site (8JE337), 1998 and 2001*. Document on file at the Florida Master Site File, Tallahassee.
Bales, Jennifer R., and Kenneth L. Kvamme
2005 Geophysical Signatures of Earthlodges in the Dakotas. In *Plains Earthlodges: Ethnographic and Archaeological Perspectives*, edited by Donna C. Roper and Elizabeth P. Pauls, pp. 157–183. University of Alabama Press, Tuscaloosa.
Barker, Alex W., Craig E. Skinner, M. Steven Shackley, Michael D. Glascock, and J. Daniel Rogers
2002 Mesoamerican Origin for an Obsidian Scraper from the Precolumbian Southeastern United States. *American Antiquity* 67:103–108.
Barrier, Casey R., and Timothy J. Horsley
2014 Shifting Communities: Demographic Profiles of Early Village Population Growth and Decline in the Central American Bottom. *American Antiquity* 79(2):295–313.
Bartington Instruments
2011a *Operation Manual for the GRAD601 Single Axis Magnetic Gradiometers*. OM1800. Bartington Instruments, Ltd., Oxford, United Kingdom.
2011b *Operation Manual for the MS2 Magnetic Susceptibility System*. Bartington Instruments, Ltd., Oxford, United Kingdom.
Basso, Keith H.
1996 *Wisdom Sits in Places: Landscape and Language among the Western Apache*. University of New Mexico Press, Albuquerque.
Beatley, Timothy, and Kristy Manning
1997 *The Ecology of Place: Planning for Environment, Economy, and Community*. Island Press, Washington, DC.
Beaudry, Mary C.
1986 The Archaeology of Historical Land Use in Massachusetts. *Historical Archaeology* 20(2):38–46.
1999 House and Household: The Archaeology of Domestic Life in Early America. In *Old and New Worlds*, edited by Geoff Egan and Ronn Michael, pp. 117–126. Oxbow Books, Oxford, United Kingdom.

Beck, Robin A., Jr., Douglas J. Bolender, James A. Brown, and Timothy K. Earle
2007 Eventful Archaeology: The Place of Space in Structural Transformation. *Current Anthropology* 48:833–860.
Bell, Robert E.
1972 *The Harlan Site, CK6, a Prehistoric Mound Center in Cherokee County, Eastern Oklahoma.* Memoir 2, Oklahoma Anthropological Society, Norman.
Beriault, John G., Robert S. Carr, Mark Lance, and Steven Bertone
2003 *A Phase I Archaeological Assessment of the Ten Thousand Islands Collier County, Florida (DHR Grant No. FO221).* AHC Technical Report No. 434. Archaeological and Historical Conservancy, Davie, Florida.
Bernbeck, Reinhard, and Randall H. McGuire (editors)
2011 *Ideologies in Archaeology.* University of Arizona Press, Tucson.
Berry, Wendell
2011 *Imagination in Place.* Counterpoint, Berkeley, California.
Bevan, Bruce W.
1983 Electromagnetics for Mapping Buried Earth Features. *Journal of Field Archaeology* 10(1):47–54.
Bigman, Daniel P.
2014 Mapping Social Relationships: Geophysical Survey of a 19th Century American Slave Cemetery. *Archaeological and Anthropological Sciences* 6:17–30.
Bigman, Daniel P., and Peter M. Lanzarone
2014 Investigating Construction History, Labour Investment, and Social Change at Ocmulgee National Monument's Mound A, Georgia, USA, Using Ground Penetrating Radar *Archaeological Prospection* 21:213–224.
Binford, Lewis Roberts
1982 The Archaeology of Place. *Journal of Anthropological Archaeology* 1:5–31.
Blitz, John H.
2010 New Perspectives in Mississippian Archaeology. *Journal of Archaeological Research* 18:1–39.
Blitz, John H., and Karl G. Lorenz
2006 *The Chattahoochee Chiefdoms.* University of Alabama Press, Tuscaloosa.
Bolender, Douglas J. (editor)
2010 *Eventful Archaeologies: New Approaches to Social Transformation in the Archaeological Record.* State University of New York Press, Albany.
Bolton, Herbert Eugene
1987 *The Hasinais: Southern Caddoans as Seen by the Earliest Europeans.* University of Oklahoma Press, Norman.
Brady, Ellen M., Victoria L. Saxe, Clay Swindell, Daniel P. Lynch, and Loretta Lautzenheiser
1998 *Archaeological Data Recovery, Site 31SP300*1* Clear Run Blacksmith-Cooper Shop, Sampson County, North Carolina.* Coastal Carolina Research, Inc., Tarboro, North Carolina. Manuscript on file, North Carolina Department of Transportation, Raleigh.
Brannan, Stefan P., and Daniel P. Bigman
2014 Ground Penetrating Radar and Resistivity Results from Mounds D and F at Singer-Moye (9SW2). *Early Georgia* 42(2):179–192.
Brown, Calvin S.
1926 *Archaeology of Mississippi.* AMS Press, Brooklyn, New York.

Brown, James A.

1996 *The Spiro Ceremonial Center: The Archaeology of Arkansas Valley Caddoan Culture in Eastern Oklahoma*. 2 vols. Memoir 29. University of Michigan Museum of Anthropology, Ann Arbor.

2010 Cosmological Layouts of Secondary Burials as Political Instruments. In *Mississippian Mortuary Practices: Beyond Hierarchy and the Representationist Perspective*, edited by Lynne P. Sullivan and Robert C. Mainfort Jr., pp. 30–53. University Press of Florida, Gainesville.

2012 Spiro Reconsidered: Sacred Economy at the Western Frontier of the Eastern Woodlands. In *The Archaeology of the Caddo*, edited by Timothy K. Perttula and Chester P. Walker, pp. 117–138. University of Nebraska Press, Lincoln.

2014 Ideological Referents of the Spiro Spirit Lodge. Paper presented at the 71st Annual Meeting of the Southeastern Archaeological Conference, Greenville, South Carolina.

Buikstra, Jane E.

1976 *Hopewell in the Lower Illinois Valley: A Regional Approach to the Study of Human Biological Variability and Prehistoric Behavior*. Northwestern University Press, Evanston, Illinois.

1981 Mortuary Site Survey in the Lower Illinois River Region (Unit VI). In *Predictive Models in Illinois Archaeology*, edited by Margaret K. Brown, pp. 73–79. Illinois Department of Conservation, Division of Historic Sites, Springfield.

1988 *The Mound-Builders of Eastern North America: A Regional Perspective*. Stichting Nederlands Museum voor Anthroplogie en Praehistorie, Amsterdam, Netherlands.

Buikstra, Jane E., and Douglas K. Charles

1999 Centering the Ancestors: Cemeteries, Mounds, and Sacred Landscapes of the Ancient North American Midcontinent. In *Archaeologies of Landscape*, edited by Wendy Ashmore and A. Bernard Knapp, pp. 201–228. Blackwell, Malden, Massachusetts.

Buikstra, Jane E., Douglas K. Charles, Kevin M. H. O'Briant, and Kyle T. Smith

1997 *The Mound House Site: Summary of the 1995–1996 Investigations Conducted by the University of Chicago, the University of New Mexico and Wesleyan University*. Center for American Archeology, Kampsville, Illinois.

Buikstra, Jane E., Douglas K. Charles, and Gordon F. M. Rakita

1998 *Staging Ritual: Hopewell Ceremonialism at the Mound House Site, Greene County, Illinois*. Kampsville Studies in Archeology and History, No. 1. Center for American Archeology, Kampsville, Illinois.

Buikstra, Jane E., Duncan P. McKinnon, Jason L. King, Jason T. Herrmann, and Mary Ann Vicari

2011 The Kampsville Experience: Education in the Lower Illinois River Valley. Poster presented at the 68th Annual Meeting of the Southeastern Archaeological Conference, Jacksonville, Florida.

Bullington, Jill

1988 Middle Woodland Mound Structure: Social Implications and Regional Context. In *The Archaic and Woodland Cemeteries at the Elizabeth Site in the Lower Illinois Valley*, edited by Douglas K. Charles, Stephen Leigh, and Jane E. Buikstra, pp. 218–241. Center for American Archeology, Kampsville, Illinois.

Burks, Jarrod, and Robert A. Cook

2011 Beyond Squier and Davis: Rediscovering Ohio's Earthworks Using Geophysical Remote Sensing. *American Antiquity* 76(4):667–689.

Canuto, Marcello A., and Jason Yaeger (editors)
2000 *The Archaeology of Communities: A New World Perspective*. Routledge, New York.
Carlson, Paul H.
1998 *The Plains Indians*. Texas A&M University Press, College Station.
Chapman, Jefferson
1985 *Tellico Archaeology: 12,000 Years of Native American History*. University of Tennessee Press, Knoxville.
Charles, Douglas K.
1985 Corporate Symbols: An Interpretive Prehistory of Indian Burial Mounds in West Central Illinois. Unpublished PhD dissertation, Department of Anthropology, Northwestern University, Evanston, Illinois.
1992 Woodland Demographic and Social Dynamics in the American Midwest: Analysis of a Burial Mound Survey. *World Archaeology* 24(2):175–197.
1995 Diachronic Regional Social Dynamics: Mortuary Sites in the Illinois Valley/American Bottom Region. In *Regional Approaches to Mortuary Analysis*, edited by Lane Anderson Beck, pp. 77–99. Plenum Press, New York.
Charles, Douglas K., and Jane E. Buikstra
2002 Siting, Sighting, and Citing the Dead. In *The Space and Place of Death*, edited by Helaine Silverman and David B. Small, pp. 13–26. Archeological Papers of the American Anthropological Association No. 11, J. K. Johnson, general editor. American Anthropological Association, Arlington, Virginia.
Charles, Douglas K., Steven R. Leigh, and Jane E. Buikstra (editors)
1988 *The Archaic and Woodland Cemeteries at the Elizabeth Site in the Lower Illinois Valley*. Center for American Archeology, Kampsville, Illinois.
Charles, Douglas K., Julieann Van Nest, and Jane E. Buikstra
2004 From the Earth: Minerals and Meaning in the Hopewellian World. In *Soils, Stones and Symbols: Cultural Perceptions of the Mineral World*, edited by Nicole Boivin and Mary Ann Owoc, pp. 43–70. UCL Press, London.
Charles, Mona
2012 *Electrical Resistance Survey of Three Sites in the Indian Camp Ranch, Montezuma County, Colorado*. Fort Lewis College, Center of Southwest Studies. Manuscript on file, Crow Canyon Archaeological Center, Cortez, Colorado.
2013 *Electrical Resistance Survey of Eight Sites in the Indian Camp Ranch, Montezuma County, Colorado*. Powderhorn Research, LLC. Manuscript on file, Crow Canyon Archaeological Center, Cortez, Colorado.
Chase, Arlen F., Diane Z. Chase, John F. Weishampel, Jason B. Drake, Ramesh L. Shrestha, K. Clint Slatton, Jaime J. Awe, and William E. Carter
2011 Airborne LiDAR, Archaeology, and the Ancient Maya Landscape at Caracol, Belize. *Journal of Archaeological Science* 38:387–398.
Chuipka, Jason P.
2009 *Animas-La Plata Project, Vol. XII: Ridges Basin Excavations: The Sacred Ridge Site*. SWCA Environmental Consultants, Phoenix, Arizona.
Clark, Anthony
1996 *Seeing Beneath the Soil: Prospecting Methods in Archaeology*. Rev. ed. B. T. Batsford, Ltd., London.
Clay, R. Berle
2001 Complementary Geophysical Survey Techniques: Why Two Ways Are Always Better Than One. *Southeastern Archaeology* 20:31–43.

2006 Conductivity Survey: A Survival Manual. In *Remote Sensing for Archaeology: An Explicitly North American Perspective*, edited by Jay K. Johnson, pp. 79–107. University of Alabama Press, Tuscaloosa.

Clements, Forrest E.

1945 Historical Sketch of the Spiro Mound. *Contributions to the Museum of the American Indian, Heye Foundation* 14:48–68.

Cobb, Charles R.

2003 Mississippian Chiefdoms: How Complex? *Annual Review of Anthropology* 32:63–85.

Cole, Fay-Cooper, and Thorne Deuel

1937 *Rediscovering Illinois: Archaeological Explorations in and around Fulton County*. University of Chicago Press, Chicago.

Collins, James M.

1997 Cahokia Settlement and Social Structure as Viewed from the Act-II. In *Cahokia: Domination and Ideology in the Mississippian World*, edited by Timothy Pauketat and Thomas E. Emerson, pp. 124–140. University of Nebraska Press, Lincoln.

Colvin, Matthew H.

2014 Assessing Monumentality in the Okeechobee Basin, Florida and the Scope of LiDAR Imaging. Paper presented at the 71st Annual Meeting of the Southeastern Archaeological Conference, Greenville, South Carolina.

Conyers, Lawrence B.

2006 Ground-Penetrating Radar. In *Remote Sensing for Archaeology: An Explicitly North American Perspective*, edited by Jay K. Johnson, pp. 131–160. University of Alabama Press, Tuscaloosa.

2010 Ground-Penetrating Radar for Anthropological Research. *Antiquity* 84(323):175–184.

2012 Advances in Ground-Penetrating Radar Exploration in Southern Arizona. *Journal of Arizona Archaeology* 3:80–91

2013 *Ground-Penetrating Radar for Archaeology*. 3rd ed. Geophysical Methods for Archaeology. AltaMira Press, Lanham, Maryland.

2014 *Interpreting Ground-Penetrating Radar for Archaeology*. Left Coast Press, Walnut Creek, California.

Conyers, Lawrence B., and Catherine M. Cameron

1998 Ground-Penetrating Radar Techniques and Three-Dimensional Computer Mapping in the American Southwest. *Journal of Field Archaeology* 25(4):417–430.

Conyers, Lawrence B., and Juerg Leckebusch

2010 Geophysical Archaeology Research Agendas for the Future: Some Ground-Penetrating Radar Examples. *Archaeological Prospection* 17:117–123.

Conyers, Lawrence B., and Jeffrey E. Lucius

1996 Velocity Analysis in Archaeological Ground-Penetrating Radar Studies. *Archaeological Prospection* 3:25–38.

Cook, Edward R., Richard Seager, Mark A. Cane, and David W. Stahle

2007 North American Drought: Reconstructions, Causes, Consequences. *Earth-Science Reviews* 81:93–134.

Cook, Robert A., and Jarrod Burks

2011 Determining Site Size and Structure: A Fort Ancient Example. *American Antiquity* 76(1):145–162.

Cook, Sherburne F., and Robert F. Heizer

1968 Relationship among Houses, Settlement Areas, and Populations in Aboriginal

California. In *Settlement Archaeology*, edited by Kwang-chih Chang, pp. 79–116. National Press Books, Palo Alto, California.

Creel, Darrell, Dale Hudler, Samuel Wilson, Chester P. Walker, and T. Clay Schultz
2008 Geophysical Survey of the Mound B Area at the George C. Davis Site. *Bulletin of the Texas Archeological Society* 79:177–190.

Cushing, Frank H.
1896 Exploration of Ancient Key Dwellers' Remains on the Gulf Coast of Florida. *Proceedings of the American Philosophical Society* 35(153):329–448.

Dalan, Rinita A.
2006 Magnetic Susceptibility. In *Remote Sensing in Archaeology: An Explicitly North American Perspective*, edited by Jay K. Johnson, pp. 161–203. University of Alabama Press, Tuscaloosa.
2008 A Review of the Role of Magnetic Susceptibility in Archaeogeophysical Studies in the USA: Recent Developments and Prospects. *Archaeological Prospection* 15(1):1–31.

Davis, Jera R., Chester P. Walker, and John H. Blitz
2015 Remote Sensing as Community Settlement Analysis at Moundville. *American Antiquity* 80(1):161–169.

Dean, Jeffrey S.
1992 Environmental Factors in the Evolution of the Chacoan Sociopolitical System. In *Anasazi Regional Organization and the Chaco System*, edited by David E. Doyel, pp. 35–44. Maxwell Museum of Anthropology Anthropological Papers No. 5. Albuquerque, New Mexico.

Dearing, J. A., R. J. L. Dann, K. Hay, J. A. Lees, P. J. Loveland, and B. A. Maher
1996 Frequency-Dependent Susceptibility Measurements of Environmental Materials. *Geophysical Journal International* 124:228–240.

Dearing, J. A., J. A. Hannam, A. S. Anderson, and E. M. H. Wellington
2001 Magnetic, Geochemical and DNA Properties of Highly Magnetic Soils in England. *Geophysical Journal International* 144:183–196.

DeBoer, Warren R.
2005 Colors for a North American Past. *World Archaeology* 37(1):66–91.

Deetz, James J. F.
1982 Households: A Structural Key to Archaeological Explanation. *American Behavioral Scientist* 25:717–724.

Delle, James A.
1998 *An Archaeology of Social Space: Analyzing Coffee Plantations in Jamaica's Blue Mountains*. Plenum Press, New York.

DeRousseau, C. Jean
1973 Mortuary Site Survey and Paleodemography in the Lower Illinois Valley. Paper presented at the 72nd Annual Meeting of the American Anthropological Association, New Orleans, Louisiana.

Deuel, Thorne
1935 Basic Cultures of the Mississippi Valley. *American Anthropologist* 37(3):429–445.

Devereux, B., G. Amable, P. Crow, and A. Cliff
2005 The Potential of Airborne LiDAR for Detection of Archaeological Features under Woodland Canopies. *Antiquity* 79:648–660.

DeVore, Steven L.
1996 *Magnetic Survey of Camp Lewis, Pecos National Historic Park, San Miguel County,*

*New Mexico*. Stewardship and Partnership Team, Rocky Mountain System Support Office, Intermountain Field Area, Denver, Colorado. Prepared for Anthropology Program, Intermountain Cultural Resource Center, Intermountain Field Area, National Park Service, Santa Fe, New Mexico.

2002 *Magnetic Susceptibility Investigations at the Barnes Folsom Site (5LA9187), Pinon Canyon Maneuver Site, Las Animas County, Colorado*. Midwest Archeological Center, National Park Service, Lincoln, Nebraska.

Diederichs, Shanna R.

2014 Electrical Resistivity Survey of Areas Adjacent to Ruins Corner (LA2520). Letter report submitted to Paleowest Archaeology, Phoenix, Arizona.

Diederichs, Shanna R., and Steven R. Copeland

2013 *The Basketmaker Communities Project Annual Report, 2012 Field Season*. Submitted to Colorado State Historic Fund. Manuscript on file at Crow Canyon Archaeological Center, Cortez, Colorado.

Diederichs, Shanna R., Steven R. Copeland, and Caitlin Sommer

2014 *The Basketmaker Communities Project Annual Report, 2013 Field Season*. Submitted to Colorado State Historic Fund. Manuscript on file at Crow Canyon Archaeological Center, Cortez, Colorado.

Dominguez, Steven

2002 Optimal Gardening Strategies: Maximizing the Input and Retention of Water in Prehistoric Gridded Fields in North Central New Mexico. *World Archaeology* 34(1):131–163.

Downum, Christian E. (editor)

2010 *Archaeological Excavations at Site B:16:105, Grand Canyon National Park*. NAU Archaeological Report 1247. Northern Arizona University, Flagstaff.

Duff, Andrew I., and Susan C. Ryan

1999 *The Shields Pueblo Research Project Annual Report, 1999 Fieldwork*. Manuscript on file, Crow Canyon Archaeological Center, Cortez, Colorado.

Effland, Richard W., Jr., A. Trinkle Jones, and Robert C. Euler

1981 *The Archaeology of Powell Plateau: Regional Interaction at Grand Canyon*. Grand Canyon Natural History Association Monograph No. 3. Grand Canyon, Arizona.

Ellison, Robert S.

1931 *Fort Bridger, Wyoming: A Brief History*. Wyoming State Archives, Museums, and Historical Department, Cheyenne.

Ernenwein, Eileen G.

2006 Imaging in the Ground-Penetrating Radar Near-Field Zone: A Case Study from New Mexico, USA. *Archaeological Prospection* 13(2):154–156.

2008 A Geophysical View of Pueblo Escondido: Implications for the Pithouse to Pueblo Transition in the Jornada Mogollon. *Bulletin of the Texas Archaeological Society* 79:125–145.

Espenshade, Chris, Terry Powis, Doug Scott, Patrick Severts, Garrett Silliman, and Sheldon Skaggs

2012 A Discussion of Standards for Metal-Detecting. Paper presented at the 2012 meeting of the Georgia Council of Professional Archaeologists, Buford.

Fairbanks, Charles H.

1977 Backyard Archaeology as a Research Strategy. In *The Conference on Historic Sites Archaeology Papers 1976, Volume 11*, edited by Stanley South, pp. 133–139. South Caro-

lina Institute of Archaeology and Anthropology, University of South Carolina, Columbia.
Fairley, Helen C.
2003 *Changing River: Time, Culture, and the Transformation of Landscape in the Grand Canyon*. SRI Technical Series 79. Tucson, Arizona.
Farnsworth, Kenneth B.
2004a *Early Hopewell Mound Explorations: The First Fifty Years in the Illinois River Valley*. Studies in Archaeology No. 3. Illinois Transportation Archaeological Research Program, University of Illinois, Urbana.
2004b Hopewellian Studies in the Lower Illinois Valley, 1878–1928. In *Early Hopewell Mound Explorations: The First Fifty Years in the Illinois River Valley*, edited by Kenneth B. Farnsworth. Studies in Archaeology No. 3. Illinois Transportation Archaeological Research Program, University of Illinois, Urbana.
Fassbinder, Jörg W. E., and Helge Stanjek
1993 Occurrence of Bacterial Magnetite in Soils from Archaeological Sites. *Archaeologia Polona* 31:117–128.
Fassbinder, Jörg W. E., Helge Stanjek, and Hojatollah Vali
1990 Occurrence of Magnetic Bacteria in Soil. *Nature* 343(6254):161–163.
Fay, Patricia M.
1987 *Archaeological Site Stabilization in the Tennessee River Valley Phase III*. Tennessee Valley Authority Publications in Anthropology No. 49. Tennessee Valley Authority, Knoxville.
Fetterman, Jerry E.
1991 Field notes regarding the great kiva at the Dillard site, 5MT10647. Documentation completed by Woods Canyon Archaeological Consultants, Yellow Jacket, Colorado. Original notes on file at Crow Canyon Archaeological Center, Cortez, Colorado.
Fewkes, Jesse Walter
1999 *Mesa Verde Ancient Architecture*. Avanyu, Albuquerque, New Mexico.
Fish, Suzanne K., Paul R. Fish, and M. Elisa Villalpando (editors)
2007 *Trincheras Sites in Time, Space, and Society*. University of Arizona Press, Tucson.
Flecher, Alice C.
1907 Kadohadacho. In *Handbook of American Indians North of Mexico*, edited by Frederick W. Hodges, pp. 638–639. Bulletin 30. Bureau of American Ethnology, Smithsonian Institution, Washington, DC.
Florida Natural Areas Inventory (FNAI)
2010 *Guide to the Natural Communities of Florida*. Florida Natural Areas Inventory, Tallahassee.
Fogelin, Lars (editor)
2008 *Religion, Archaeology, and the Material World*. Occasional Paper No. 36. Center for Archaeological Investigations, Southern Illinois University, Carbondale.
Fowler, Melvin L.
1997 *The Cahokia Atlas: A Historical Atlas of Cahokia Archaeology*. Illinois Transportation Archaeological Research Program, University of Illinois, Urbana.
Gaffin, Mary Lee, Keith J. Little, Hunter B. Johnson, and Ann Marshall
2012 *An Archaeological Survey of the Raccoon Creek Management Area in the Guntersville Basin, Jackson County, Alabama*. Tennessee Valley Archaeological Research, Huntsville, Alabama.

Gaffney, Chris, Vince Gaffney, Wolfgang Neubauer, Eamonn Baldwin, Henry Chapman, Paul Garwood, Helen Moulden, Tom Sparrow, Richard Bates, Klaus Löcker, Alois Hinterleitner, Immo Trinks, Erich Nau, Thomas Zitz, Sebastian Floery, Geert Verhoeven, and Michael Doneus

2012 The Stonehenge Hidden Landscapes Project. *Archaeological Prospection* 19:147–155.

Gaffney, Chris F., and John A. Gater

2003 *Revealing the Buried Past: Geophysics for Archaeologists.* Tempus, Stroud, Gloucestershire, England.

Gaffney, Chris F., John A. Gater, Paul Linford, Vince L. Gaffney, and Roger White

2000 Large-Scale Systematic Fluxgate Gradiometry at the Roman City of Wroxeter. *Archaeological Prospection* 7:81–99.

Gaffney, Chris, Gwilym Hughes, and John Gater

2005 Geophysical Surveys at King Lobengula's Palace KoBulawayo, Zimbabwe. *Archaeological Prospection* 12:31–49.

Gasco, Janine

1992 Documentary and Archaeological Evidence of Household Differentiation in Colonial Soconusco, New Spain. In *Text-Aided Archaeology*, edited by Barbara J. Little, pp. 83–96. CRC Press, Boca Raton, Florida.

Girard, Jeffrey S.

2012 Settlement Patterns and Variation in Caddo Pottery Decoration: A Case Study of the Willow Chute Bayou Locality. In *The Archaeology of the Caddo*, edited by Timothy K. Perttula and Chester P. Walker, pp. 239–287. University of Nebraska Press, Lincoln.

Glowacki, Mary

2010 National Register of Historic Places registration form for Letchworth Mounds Archaeological Site. Document on file at the Florida Master Site File, Tallahassee.

Goodman, Dean

2004 GPR Survey at the Spiro Mounds, Oklahoma. Unpublished manuscript on file at Spiro Mounds Archaeological Center, Spiro, Oklahoma.

Goodman, Dean, Yasushi Nishimura, and J. Daniel Rogers

1995 GPR (Ground Penetrating Radar) Time Slices in Archaeological Prospection. *Archaeological Prospection* 2:85–89.

Goodman, Dean, and Salvatore Piro

2013 *GPR Remote Sensing in Archaeology.* Geotechnologies and the Environment 9. Springer, Heidelberg, Germany.

Gorin, Betty J.

2006 *Morgan Is Coming! Confederate Raiders in the Heartland of Kentucky.* Harmony House, Louisville, Kentucky.

GRASS Development Team

2012 *Geographic Resources Analysis Support System (GRASS GIS) Software.* Open Source Geospatial Foundation, USA.

Gremillion, Kristen J.

2002 The Development and Dispersal of Agricultural Systems in the Woodland Period Southeast. In *The Woodland Southeast*, edited by David G. Anderson and Robert C. Mainfort Jr., pp. 483–501. University of Alabama Press, Tuscaloosa.

Griffin, James B., and James A. Ford

1940 Site card on file with Mississippi Department of Archives and History, Jackson.

Griffin, John W.

2002 *Archaeology of the Everglades.* University Press of Florida, Gainesville.

Griffith, William
1954 *The Hasinai Indians of East Texas as Seen by Europeans, 1687–1772*. Middle American Research Institute, Tulane University, New Orleans, Louisiana.
Hackbarth, M. R.
2012 *Phase II Archaeological Data Recovery in the Palo Verde Ruin Open Space Park, AZ T:8:68 (ASM) Phoenix, Maricopa County, Arizona*. Report on file at the Arizona State Museum, Tucson.
Halbwachs, Maurice
1992 *On Collective Memory*. Translated by Lewis A. Coser. University of Chicago Press, Chicago.
Haley, Bryan S.
2002 Airborne Remote Sensing, Image Processing, and Multisensor Data Fusion at the Hollywood Site, a Large Late Mississippian Mound Center. Unpublished master's thesis, Department of Sociology and Anthropology, University of Mississippi, Oxford.
2014a The Big Picture at Hollywood: Geophysical and Archaeological Investigations at a Mississippian Mound Centre. *Archaeological Prospection* 21:39–47.
2014b An Updated Chronology for the Hollywood Site. Paper presented at the 2014 meeting of the Mississippi Archaeological Association, Jackson.
Hall, Robert L.
1979 In Search of the Ideology of the Adena-Hopewell Climax. In *Hopewell Archaeology: The Chillicothe Conference*, edited by David S. Brose and N'omi Greber, pp. 258–265. Kent State University Press, Kent, Ohio.
1997 *An Archaeology of the Soul: North American Indian Belief and Ritual*. University of Illinois Press, Urbana.
Hally, David J.
1996 Platform Mound Construction and the Instability of Mississippian Chiefdoms. In *Political Structure and Change in the Prehistoric Southeastern United States*, edited by John F. Scarry, pp. 92–127. University Press of Florida, Gainesville.
2008 *King: The Social Archaeology of a Late Mississippian Town in Northwestern Georgia*. University of Alabama Press, Tuscaloosa.
Hally, David J., and Hypatia Kelly
1998 The Nature of Mississippian Towns in Georgia: The King Site Example. In *Mississippian Towns and Sacred Spaces: Searching for an Architectural Grammar*, edited by R. Barry Lewis and Charles B. Stout, pp. 49–63. University of Alabama Press, Tuscaloosa.
Hamilton, Henry W.
1952 The Spiro Mound. *Missouri Archaeologist* 14:17–88.
Hammerstedt, Scott W., Amanda L. Regnier, and Patrick C. Livingood
2010 Geophysical and Archaeological Investigations at the Clement Site, a Caddo Mound Complex in Southeastern Oklahoma. *Southeastern Archaeology* 29:279–291.
Hargrave, Michael L.
2006 Ground Truthing the Results of Geophysical Surveys. In *Remote Sensing in Archaeology: An Explicitly North American Perspective*, edited by Jay K. Johnson, pp. 269–304. University of Alabama Press, Tuscaloosa.
Hargrave, Michael L., Tad Britt, and Matthew D. Reynolds
2007 Magnetic Evidence of Ridge Construction and Use at Poverty Point. *American Antiquity* 72(4):757–770.

Harmon, James M., Mark P. Leone, Stephen D. Prince, and Marcia Snyder

2006 LiDAR for Archaeological Landscape Analysis: A Case Study of Two Eighteenth-Century Maryland Plantation Sites. *American Antiquity* 71(4):649–670.

Harmon, Michael A., and Rodney J. Snedeker

1997 The Archaeological Record of Tar and Pitch Production in Coastal Carolina. In *Carolina's Historical Landscapes: Archaeological Perspectives*, edited by Linda F. Stine. University of Tennessee Press, Knoxville.

Heimmer, Don H., and Steven L. De Vore

1995 *Near-Surface, High Resolution Geophysical Methods for Cultural Resource Management and Archaeological Investigations*. Rev. ed. US Department of the Interior, National Park Service, Denver, Colorado.

Henderson, Amy

2004 *The ParaView Guide*. Kitware, Clifton Park, New York.

Henry, Edward R.

2011 A Multistage Geophysical Approach to Detecting and Interpreting Archaeological Features at the LeBus Circle, Bourbon County, Kentucky. *Archaeological Prospection* 18:231–244.

Henry, Edward R., Nicolas R. Laracuente, Jared S. Case, and Jay K. Johnson

2014 Incorporating Multistaged Geophysical Survey into Regional-Scale Models: A Case Study from an Adena Burial Mound in Central Kentucky. *Archaeological Prospection* 21:15–26.

Herrmann, Jason T., Jason L. King, and Jane E. Buikstra

2014 Mapping the Internal Structure of Hopewell Tumuli in the Lower Illinois River Valley through Archaeological Geophysics. *Advances in Archaeological Practice: A Journal of the Society for American Archaeology* 2(3):164–179.

Hinz, Emily A., John F. Ferguson, Louise Pellerin, and Ann F. Ramenofsky

2008 A Geophysical Investigation of Subsurface Structures and Quaternary Geology at San Marcos Pueblo, New Mexico. *Archaeological Prospection* 15(4):247–265.

Hoagland, Alison K.

2001 Architecture and Interpretation at Forts Laramie and Bridger. *Public Historian* 23(1):27–54.

Hoffman, Michael

1970 Archaeological and Historical Assessment of the Red River Basin in Arkansas. In *Archeological and Historical Resources of the Red River Basin*, edited by Hester A. Davis, pp. 137–194. Research Series No. 1. Arkansas Archeological Survey, Fayetteville.

Holliday, Steven E.

1977 Locational Analysis and Mortuary Site Distribution in the Lower Illinois River Valley. Master's thesis, Department of Anthropology, Northwestern University, Evanston, Illinois.

Horn, Jonathon C., Jerry Fetterman, and Linda Honeycutt (compilers)

2003 *The Mid-America Pipeline Company/Williams Rocky Mountain Expansion Loop Pipeline Archaeological Data Recovery Project, Northwestern New Mexico, Western Colorado, and Eastern Utah*. Submitted to Williams Companies, Inc., Tulsa, Oklahoma. Report on file at the Colorado Office of Archaeology and Historic Preservation, Denver.

Howard, James H.

1962 *Report of the Investigations of the Huff Site, 32MO11, Morton County, North Dakota, 1959*. University of North Dakota Anthropological Papers No. 2. Grand Forks, North Dakota.

Howard, Lynn H.
1948 Battle Mound daily log notes. Notes on file, Arkansas Archeological Survey, Fayetteville.

Hudson, Charles M.
1976 *The Southeastern Indians*. University of Tennessee Press, Knoxville.

Hyde, George E.
1974 *The Pawnee Indians*. University of Oklahoma Press, Norman.

Jackson, Edwin H., Susan L. Scott, and Frank F. Schambach
2012 At the House of the Priest: Faunal Remains from the Crenshaw Site (3MI6), Southwest Arkansas. In *The Archaeology of the Caddo*, edited by Timothy K. Perttula and Chester P. Walker, pp. 47–85. University of Nebraska Press, Lincoln.

Jensen, John R.
2005 *Introductory Digital Image Processing: A Remote Sensing Perspective*. 3rd ed. Pearson Prentice Hall, Upper Saddle River, New Jersey.

Jeter, Marvin D. (editor)
2009 *Edward Palmer's Arkansaw Mounds*. University of Alabama Press, Tuscaloosa.

Johannessen, Sissel
1993 Farmers of the Late Woodland. In *Foraging and Farming in the Eastern Woodlands*, edited by Margaret Scarry, pp. 57–77. University Press of Florida, Gainesville.

Johnson, Jay K.
2005 A Structural Comparison of Two Late Mississippian Mound Centers in the Yazoo Basin, Mississippi. Paper presented at the 2005 Southeastern Archaeological Conference, Columbia, South Carolina.

Johnson, Jay K. (editor)
2006 *Remote Sensing in Archaeology: An Explicitly North American Perspective*. University of Alabama Press, Tuscaloosa.

Johnson, Jay K., Richard Stallings, Nancy Ross-Stallings, R. Berle Clay, and V. Stephen Jones
2000 *Remote Sensing and Ground Truth at the Hollywood Mounds Site in Tunica County, Mississippi*. Report prepared for the Mississippi Department of Archives and History, House Bill 1502, Grant Project GS#501–074.

Johnson, Katharine M., and William B. Ouimet
2014 Rediscovering the Lost Archaeological Landscape of Southern New England Using Airborne Light Detection and Ranging (LiDAR). *Journal of Archaeological Science* 43:9–20.

Johnson, William J.
2003 *Geophysical Survey of LA 1674, Aztec Ruins National Monument, D'Appolonia, Monroeville, Pennsylvania*. Submitted to Aztec Ruins National Monument.

Johnston, Richard B.
1965 Trials of the ELSEC Proton Magnetometer at Two Sites on Wetherill Mesa. In *Contributions of the Wetherill Mesa Archaeological Project*, assembled by Douglas Osborne, pp. 180–181. Society for American Archaeology Memoir No. 19, Washington DC.

Jones, B. Calvin, Daniel T. Penton, and Louis D. Tesar
1998 1973 and 1994 Excavations at the Block-Sterns Site, Leon County, Florida. In *A World Engraved: Archaeology of the Swift Creek Culture*, edited by M. Williams and Daniel T. Elliot, pp. 222–246. University of Alabama Press, Tuscaloosa.

Joseph, Joe W., and Bradford Y. Botwick
2009 *Rural Industries of the Sand Hills, Georgia, South Carolina and North Carolina,*

*a Historic Context*. New South Associates Technical Report 1799. Public version, Legacy Resource Management Program, Department of Defense, Legacy Project No. 09–436.

Joseph, Joe W., Theresa M. Hamby, and Catherine S. Long (editors)
2004 *Historical Archaeology in Georgia*. University of Georgia Laboratory of Archaeology Series Report No. 39, Georgia Archaeological Research Design Paper No. 14.

Judge, W. James
1991 Chaco: Current Views of Prehistory and the Regional System. In *Chaco and Hohokam: Prehistoric Regional Systems in the American Southwest*, edited by P. L. Crown and J. Judge, pp. 11–30. School of American Research Press, Santa Fe, New Mexico.

Kassabaum, Megan C., Edward R. Henry, Vincas P. Steponaitis, and John W. O'Hear
2014 Between Surface and Summit: The Process of Mound Construction at Feltus. *Archaeological Prospection* 21:27–37.

Kay, Marvin, and George Sabo III
2006 Mortuary Ritual and Winter Solstice Imagery of the Harlan-Style Charnel House. *Southeastern Archaeology* 25:29–47.

Kelley, David B. (editor)
1997 *Two Caddoan Farmsteads in the Red River Valley*. Arkansas Archeological Survey Research Series No. 51. Fayetteville, Arkansas.

Kidder, T. R.
2004 Plazas as Architecture: An Example from the Raffman Site, Northeast Louisiana. *American Antiquity* 69:514–532.
2010 Hunter-Gatherer Ritual and Complexity: New Evidence from Poverty Point, Louisiana. In *Ancient Complexities: New Perspectives in Precolumbian North America*, edited by Susan M. Alt, pp. 32–51. University of Utah Press, Salt Lake City.
2011 Transforming Hunter-Gatherer History at Poverty Point. In *Hunter-Gatherer Archaeology as Historical Process*, edited by Kenneth E. Sassaman and Donald H. Holley Jr., pp. 95–119. University of Arizona Press, Tucson.

King, Adam
2003 *Etowah: The Political History of a Chiefdom Capital*. University of Alabama Press, Tuscaloosa.

King, Adam, and Maureen S. Meyers
2002 Exploring the Edges of the Mississippian World. *Southeastern Archaeology* 21(2): 113–116.

King, Adam, Chester P. Walker, Robert V. Sharp, F. Kent Reilly, and Duncan P. McKinnon
2011 Remote Sensing Data from Etowah's Mound A: Architecture and the Re-creation of Mississippian Tradition. *American Antiquity* 76(2):355–371.

King, Jason L., and Jane E. Buikstra
2010 Liturgic Sequences and Variability in Middle Woodland Floodplain Mounds. Paper presented at the 75th Annual Meeting of the Society for American Archaeology, Saint Louis, Missouri.

King, Jason L., Jane E. Buikstra, and Douglas K. Charles
2011 Time and Archaeological Traditions in the Lower Illinois Valley. *American Antiquity* 76(3):500–528.

Kitchens, Hershel, and Dot Kitchens
1968 Original unpublished field notes from Battle Mound (3LA1). Manuscripts on file, Arkansas Archeological Survey, Fayetteville.

Knight, Vernon James, Jr.

1986 The Institutional Organization of Mississippian Religion. *American Antiquity* 51:675–687.

1998 Moundville as a Diagrammatic Ceremonial Center. In *Archaeology of the Moundville Chiefdom*, edited by Vernon James Knight Jr. and Vincas P. Steponaitis, pp. 44–62. University of Alabama Press, Tuscaloosa.

2007 Conclusions: Taking Architecture Seriously. In *Architectural Variability in the Southeast*, edited by Cameron H. Lacquement, pp. 186–192. University of Alabama Press, Tuscaloosa.

2010 *Mound Excavations at Moundville: Architecture, Elites, and Social Order*. University of Alabama Press, Tuscaloosa.

Knight, Vernon James, Jr., and Vincas P. Steponaitis

2006 *Archaeology of the Moundville Chiefdom: Chronology, Content, Contest*. University of Alabama Press, Tuscaloosa.

Koerner, Shannon D.

2012 Revisiting Mississippian Origins in Eastern Tennessee. Paper Presented at the 69th Annual Meeting of the Southeastern Archaeological Conference, Baton Rouge, Louisiana.

Koerner, Shannon D., Lynne P. Sullivan, and Stephen J. Yerka

2010 *Archaeological Investigations at Site 40KN266, a Mississippian Period Town on the French Broad River, Knox County, Tennessee*. Frank H. McClung Museum and the Archaeological Research Laboratory, Department of Anthropology, University of Tennessee, Knoxville.

Kohler, Timothy A., Matt Pier Glaude, Jean-Pierre Bocquet-Appel, and Brian M. Kemp

2008 The Neolithic Demographic Transition in the U.S. Southwest. *American Antiquity* 74(4):645–669.

Kohler, Timothy A., and Charles Reed

2011 Explaining the Structure and Timing of Formation of Pueblo I Villages in the Northern U.S. Southwest. In *Sustainable Lifeways: Cultural Persistence in an Ever-Changing Environment*, edited by Naomi F. Miller, Katherine M. Moore, and Kathleen Ryan. University of Pennsylvania Museum of Archaeology and Anthropology, Philadelphia.

Kornfeld, Marcel, and Alan J. Osborn (editors)

2003 *Islands on the Plains: Ecological, Social, and Ritual Use of Landscapes*. University of Utah Press, Salt Lake City.

Kozuch, Laura

2002 Olivella Beads from Spiro and the Plains. *American Antiquity* 67:697–709.

Krieger, Alex D.

1949 *Viking Fund Grant for Battle Mound Excavations, Lafayette County, Arkansas: A Report of Activities and Expenditures as of January 31, 1949*. Manuscript on file, Arkansas Archeological Survey, Fayetteville.

Kroll, Ellen M., and T. Douglas Price (editors)

1991 *The Interpretation of Archaeological Spatial Patterning*. Springer, New York.

Kruse, Fred A., and Sandra L. Perry

2013 Mineral Mapping Using Simulated Worldview-3 Short-Wave-Infrared Imagery. *Remote Sensing* 5:2688–2703.

Kujit, Ian

1996 Negotiating Equality through Ritual: A Consideration of Late Natufian and Prepottery Neolithic. *Journal of Anthropological Archaeology* 14(4):361–440.

Kvamme, Kenneth L.
2003a Geophysical Surveys as Landscape Archaeology. *American Antiquity* 68:435–457.
2003b Multidimensional Prospecting in North American Great Plains Village Sites. *Archaeological Prospection* 10:131–142.
2006a Data Processing and Presentation. In *Remote Sensing in Archaeology: An Explicitly North American Perspective*, edited by Jay K. Johnson, pp. 235–250. University of Alabama Press, Tuscaloosa.
2006b Magnetometry: Nature's Gift to Archaeology. In *Remote Sensing in Archaeology: An Explicitly North American Perspective*, edited by Jay K. Johnson, pp. 205–233. University of Alabama Press, Tuscaloosa.
2007a Geophysical Mappings and Findings in Northern Plains Village Sites. In *Plains Village Archaeology: Bison-Hunting Farmers in the Central and Northern Plains*, edited by Stanley A. Ahler and Marvin Kay, pp. 210–221. University of Utah Press, Salt Lake City.
2007b Integrating Multiple High Resolution Geophysical Data Sets. In *Remote Sensing in Archaeology*, edited by J. Wiseman and F. El-Baz, pp. 146–163. Plenum Press, New York.
2008a *Geophysical Investigations and Topographic Mapping during 2007 at the Larson Site (32BL9), North Dakota*. Report submitted to PaleoCultural Research Group, Flagstaff, Arizona.
2008b Remote Sensing Approaches to Archaeological Reasoning: Pattern Recognition and Physical Principles. In *Archaeological Concepts for the Study of the Cultural Past*, edited by Alan P. Sullivan III, pp. 65–84. University of Utah Press, Salt Lake City.
Kvamme, Kenneth L., and Stanley A. Ahler
2000 Remote Sensing and Related Investigations. In *New Geophysical and Archaeological Investigations at Huff Village State Historic Site (32MO11), Morton County, North Dakota*, edited by Stanley A. Ahler and Kenneth L. Kvamme, pp. 11–56. Research Contribution No. 26. PaleoCultural Research Group, Flagstaff, Arizona. Submitted to State Historical Society of North Dakota, Bismarck.
2007 Integrated Remote Sensing and Excavation at Double Ditch State Historic Site, North Dakota. *American Antiquity* 72(3):539–561.
Kvamme, Kenneth L., Eileen G. Ernenwein, Michael L. Hargrave, Thomas Sever, Deborah Harmon, and Fred Limp
2006 *New Approaches to the Use and Integration of Multi-Sensor Remote Sensing for Historic Resource Identification and Evaluation*. Center for Advanced Spatial Technologies, University of Arkansas, Fayetteville.
Kvamme, Kenneth L., Eileen G. Ernenwein, and Christine J. Markussen
2006 Robotic Total Station for Microtopographic Mapping: An Example from the Northern Great Plains. *Archaeological Prospection* 13:91–102.
Kvamme, Kenneth L., Duncan P. McKinnon, and Adam S. Wiewel
2009 *Geophysical Investigations and Topographic Mapping during 2009 at Huff Village (32MO11), North Dakota*. Archaeo-Imaging Lab, Department of Anthropology, University of Arkansas, Fayetteville. Submitted to the State Historical Society of North Dakota, Bismarck.
Kvamme, Kenneth L., and Adam Wiewel
2010 *Final Report of Geophysical Investigations at 10EL462, near Mountain Home, Idaho*. Report submitted to Idaho Power Company, Boise.

Lacquement, Cameron (editor)
2007 *Architectural Variability in the Southeast*. University of Alabama Press, Tuscaloosa.
Lankford, George E., F. Kent Reilly III, and James F. Garber
2011 *Visualizing the Sacred: Cosmic Visions, Regionalism, and the Art of the Mississippian World*. University of Texas Press, Austin.
La Vere, David
2007 *Looting Spiro Mounds: An American King Tut's Tomb*. University of Oklahoma Press, Norman.
Le Borgne, E.
1955 Susceptibilité magnétique anormale du sol superficiel. *Annales de Géophysique* 11:399–419.
1960 Influence du feu sur les propriétés magnétiques du sol et sur celles du schiste et du granite. *Annales de Géophysique* 16:159–195.
Lehmer, Donald J.
1971 *Introduction to Middle Missouri Archeology*. National Park Service, Anthropological Papers No. 1. Department of the Interior, Washington, DC.
Lekson, Stephen H.
2007 *The Architecture of Chaco Canyon, New Mexico*. University of Utah Press, Salt Lake City.
Lengyel, Stacey
2013 *Archaeomagnetic Laboratory Report*. Archaeomagnetic Laboratory, Illinois State Museum, Springfield. Manuscript on file, Crow Canyon Archaeological Center, Cortez, Colorado.
Le Page du Pratz, Antoine
1975 The History of Louisiana. Facsimile reproduction of the 1774 edition, edited by Joseph G. Tregle Jr. Louisiana State University Press, Baton Rouge.
Lewis, R. Barry, and Charles B. Stout (editors)
1998 *Mississippian Towns and Sacred Spaces: Searching for an Architectural Grammar*. University of Alabama Press, Tuscaloosa.
Lewis, Thomas M. N., and Madeline D. Kneberg
1946 *Hiwassee Island: An Archaeological Account of Four Tennessee Indian Peoples*. University of Tennessee Press, Knoxville.
Lewis, Thomas M. N., and Madeline D. Kneberg Lewis
1995 *The Prehistory of the Chickamauga Basin in Tennessee*, edited by Lynne P. Sullivan. 2 vols. University of Tennessee Press, Knoxville.
Lightfoot, Ricky R.
1994 *The Duckfoot Site, Volume 2: Archaeology of the House and Household*. Occasional Papers No. 4. Crow Canyon Archaeological Center, Cortez, Colorado.
Lindauer, Owen, and John H. Blitz
1997 Higher Ground: The Archaeology of North American Platform Mounds. *Journal of Archaeological Research* 5(2):169–207.
Linford, Neal T.
2004 Magnetic Ghosts: Mineral Magnetic Measurements on Roman and Anglo-Saxon Graves. *Archaeological Prospection* 11(3):167–180.
Linton, Ralph
1923 *Purification of the Sacred Bundles, a Ceremony of the Pawnee*. Fieldiana Popular Series, Anthropology No. 7. Field Museum of Natural History, Chicago.

Little, Barbara J., and Paul A. Shackel (editors)
2007 *Archaeology as a Tool of Civic Engagement*. AltaMira Press, Lanham, Maryland.
Liu, Quingsong, Andrew P. Roberts, Juan C. Larrasoaña, Subir K. Banerjee, Yohan Guyodo, Lisa Tauxe, and Frank Oldfield
2012 Environmental Magnetism: Principles and Applications. *Reviews of Geophysics* 50(4):RG4002.
Lockhart, Jami J.
2007 Prehistoric Caddo of Arkansas: A Multiscalar Examination of Past Cultural Landscapes. PhD dissertation, Environmental Dynamics Program, University of Arkansas, Fayetteville.
2012 Spatial Patterns of Caddo Mound Sites in the West Gulf Coastal Plain of Arkansas. In *The Archaeology of the Caddo*, edited by Timothy K. Perttula and Chester P. Walker, pp. 313–334. University of Nebraska Press, Lincoln.
Lockhart, Jami J., and Thomas Green
2006 The Current and Potential Role of Archaeogeophysics in Cultural Resource Management in the United States. In *Remote Sensing in Archaeology: An Explicitly North American Perspective*, edited by Jay K. Johnson, pp. 17–32. University of Alabama Press, Tuscaloosa.
Loose, Richard W., and Thomas R. Lyons
1976 The Chetro Ketl Field: A Planned Water Control System in Chaco Canyon. In *Remote Sensing Experiments in Cultural Resource Studies: Non-destructive Methods of Archeological Exploration, Survey, and Analysis*. Reports of the Chaco Center 1:133–153. National Park Service and University of New Mexico, Albuquerque.
Love, David W.
1980 Quaternary Geology of Chaco Canyon, Northwestern New Mexico. PhD dissertation, University of New Mexico, Albuquerque.
Luby, Edward M., Clayton D. Drescher, and Kent G. Lightfoot
2006 Shell Mounds and Mounded Landscapes in the San Francisco Bay Area: An Integrated Approach. *Journal of Island and Coastal Archaeology* 1:191–214.
Luer, George M.
1989 Calusa Canals in Southwestern Florida: Routes of Tribute and Exchange. *Florida Anthropologist* 42:89–130.
Lynch, David, and Rory Becker
2008 *2007 Geophysical Survey of the Fort Bridger State Historic Site (48(UT)29)*. Manuscript on file at Fort Bridger State Historic Site, Fort Bridger, Wyoming.
Lyons, Thomas R., and Thomas E. Avery
1977 *Remote Sensing: A Handbook for Archaeologists and Cultural Resource Managers*. Cultural Resources Management, National Park Service, US Department of the Interior, Washington, DC.
Maher, Barbara A., and Reginald M. Taylor
1988 Formation of Ultrafine-Grained Magnetite in Soils. *Nature* 336(6197):368–370.
Maki, David, Jeffrey A. Homburg, and Scott D. Brosowske
2006 Thermally Activated Mineralogical Transformations in Archaeological Hearths: Inversion from Maghemite γFe2O4 Phase to Haematite αFe2O4 Form. *Archaeological Prospection* 13(3):207–227.
Marquardt, William H.
2010a Mounds, Middens, and Rapid Climate Change during the Archaic-Woodland Transition in the Southeastern United States. In *Trend, Tradition, and Turmoil: What*

*Happened to the Southeastern Archaic*, edited by David Hurst Thomas and Matthew C. Sanger, pp. 253–272. Anthropological Papers of the American Museum of Natural History Vol. 93, New York.

2010b Shell Mounds in the Southeast: Middens, Monuments, Temple Mounds, Rings, or Works? *American Antiquity* 75:551–570.

Marquardt, William H. (editor)

1992 *Culture and Environment in the Domain of the Calusa*. Monograph No. 1, Institute of Archaeology and Paleoenvironmental Studies. University Press of Florida, Gainesville.

Marquardt, William H., and Karen J. Walker

2012 Southwest Florida during the Mississippian Period. In *Late Prehistoric Florida: Archaeology at the Edge of the Mississippian World*, edited by Keith H. Ashley and Nancy M. White, pp 29–61. University Press of Florida, Gainesville.

2013 The Pineland Site Complex: An Environmental and Cultural History. In *The Archaeology of Pineland: A Coastal Southwest Florida Site Complex, A.D. 50–1710*, edited by William H. Marquardt and Karen J. Walker, pp. 793–920. Monograph No. 4, Institute of Archaeology and Paleoenvironmental Studies. University Press of Florida, Gainesville.

Marrinan, Rochelle A., and Nancy Marie White

2007 Modeling Fort Walton Culture in Northwest Florida. *Southeastern Archaeology* 26:292–318.

Martin, Glenn

1939 *Notes on Certain Explorations on the Battle Place, Red River Bottoms, Lafayette County, Arkansas*. Manuscript on file, Gilcrease Museum, Tulsa, Oklahoma.

Martin, John F.

1985 The Prehistory and Ethnohistory of Havasupai-Hualapai Relations. *Ethnohistory* 31:135–153.

Maruca, Mary V.

1980 Remote Sensing: Tool for the Park Manager. *Courier: The National Park Service Newsletter* 3(11):1–3.

Mathien, Frances J.

2005 *Cultural Ecology of Chaco Canyon and the San Juan Basin*. Publications in Archaeology 18H, Chaco Canyon Studies. National Park Service, Santa Fe, New Mexico.

Maxwell, Timothy D., and Kurt F. Anschuetz

1992 The Southwestern Ethnographic Record and Prehistoric Agricultural Diversity. In *Gardens of Prehistory: The Archaeology of Settlement Agriculture in Greater Mesoamerica*, edited by Thomas W. Killion, pp. 35–68. University of Alabama Press, Tuscaloosa.

McAdams, William, Jr.

1881 Ancient Mounds of Illinois. *Proceedings of the National Academy of Science*. Salem, Massachusetts.

1884 Mounds of the Mississippi Bottom, Illinois. In *Annual Report (1882) of the Smithsonian Institution*, pp. 684–686. Smithsonian Institution, Government Printing Office, Washington, DC.

1887 *Records of Ancient Races in the Mississippi Valley*. CR Barns, Saint Louis.

McCoy, Ronald

1990 The Blue Green Water People: The Havasupai of the Grand Canyon. *The World & I Journal* 5(3):619–629.

McKern, William C.
1939 The Midwestern Taxonomic Method as an Aid to Archaeological Culture Study. *American Antiquity* 4(4):301–313.
McKinnon, Duncan P.
2009 Exploring Settlement Patterning at a Premier Caddo Mound Site in the Red River Great Bend Region. *Southeastern Archaeology* 28(2):248–258.
2010a Summer 1948: A Summary of the Krieger Excavations at Battle Mound (3LA1), a Premier Caddo Mound Site in the Red River Great Bend Region. *Arkansas Archeologist* 49:1–16.
2010b Continuing the Research: Archaeogeophysical Investigations at the Battle Mound Site (3LA1) in Lafayette County, Arkansas. *Southeastern Archaeology* 29(2):250–260.
2011 Thirty Years in the Foil: An ARF-Funded AMS Date from Battle Mound (3LA1). *Field Notes: Newsletter of the Arkansas Archeological Society* 362:6–8.
2012 Corn Cobs Collected in 1948: An ARF-Funded AMS Date from Battle Mound (3LA1). *Field Notes: Newsletter of the Arkansas Archeological Society* 368:8–9.
2013a Battle Mound: Exploring Space, Place, and History of a Red River Caddo Community in Southwest Arkansas. PhD dissertation, Department of Anthropology, University of Arkansas, Fayetteville.
2013b Landscape as a Ritual Object: Exploring Some Thoughts on Organized Space in the Great Bend Region in Southwestern Arkansas. *Caddo Archeology Journal* 23:67–84.
2013c Making Headway: Another ARF-Funded AMS Date from Battle Mound (3LA1). *Field Notes: Newsletter of the Arkansas Archeological Society* 372:9–10.
2017 *The Battle Mound Landscape: Exploring Space, Place, and History of a Red River Caddo Community in Southwest Arkansas*. Research Series No. 68. Arkansas Archeological Society, Fayetteville.
McKinnon, Duncan P., Jason L. King, Jane E. Buikstra, Taylor H. Thornton, and Jason T. Herrmann
2016 Returning to the Kamp Mound Group (11C12): Results from Geomagnetic Survey and High-Density Topographic Mapping in Calhoun County, Illinois. *Midcontenental Journal of Archaeology* 41(3): 231–254.
McKinnon, Duncan P., and Kenneth L. Kvamme
2007 Geophysical Investigations at the National Historical Landmark Site of Ft. Pierre Chouteau. Paper presented at the 65th Annual Meeting of the Plains Anthropological Conference, Rapid City, South Dakota.
McMullan, Robin Carol
2001 Lake Mound Plantation: Historic Occupations of Toltec Mounds State Park. Master's thesis, Department of Anthropology, University of Arkansas, Fayetteville.
McNiven, Ian J.
2012 Ritualized Middening Practices. *Journal of Archaeological Method and Theory* 20: 552–587.
Mehta, Jayur Madhusudan
2007 A Study of Sweat Lodges in the Southeastern United States. Unpublished master's thesis, Department of Anthropology, University of Alabama, Tuscaloosa.
Milanich, Jerald T.
1994 *Archaeology of Precolumbian Florida*. University Press of Florida, Gainesville.
2002 Weeden Island Cultures. In *The Woodland Southeast*, edited by David G. Anderson and Robert C. Mainfort Jr., pp. 352–372. University of Alabama Press, Tuscaloosa.

Milanich, Jerald T., Ann S. Cordell, Vernon James Knight, Timothy A. Kohler, and Brenda J. Sigler-Lavelle
1997 *Archaeology of Northern Florida A.D. 200–900: The McKeithen Weeden Island Culture*. University Press of Florida, Gainesville.
Milner, George R.
1998 *The Cahokia Chiefdom: The Archaeology of a Mississippian Society*. Smithsonian Institution Press, Washington, DC.
2004 Old Mounds, Ancient Hunter-Gatherers, and Modern Archaeologists. In *Signs of Power: The Rise of Cultural Complexity in the Southeast*, edited by Jon L. Gibson and Philip J. Carr, pp. 300–315. University of Alabama Press, Tuscaloosa.
Mink, Philip B., II
2013 *Remote Sensing Five Archaeological Sites along the Colorado Rive*r. Report on file at Grand Canyon National Park, Grand Canyon, Arizona.
2015 Living on the Edge: Rethinking Pueblo Period (AD 700–AD 1225) Indigenous Settlement Patterns within Grand Canyon National Park. Unpublished PhD dissertation, Department of Anthropology, University of Kentucky, Lexington.
Mitchell, Mark D.
2013 *Crafting History in the Northern Plains: A Political Economy of the Heart River Region, 1400–1750*. University of Arizona Press, Tucson.
Monaghan, G. William, and Christopher S. Peebles
2010 The Construction, Use and Abandonment of Angel Site Mound A: Tracing the History of a Middle Mississippian Town through its Earthworks. *American Antiquity* 75:935–953.
Moore, Clarence B.
1912 Some Aboriginal Sites on the Red River. *Journal of the Academy of Natural Sciences of Philadelphia* 14:481–638.
1915 Aboriginal Sites on the Tennessee River. *Journal of the Academy of Natural Sciences of Philadelphia* 16:171–422.
Morain, Stanley A., Thomas K. Budge, and Amelia Komarek
1981 *Remote Sensing: Multispectral Analyses of Cultural Resources, Supplement No. 5*. Cultural Resources Management, National Park Service, US Department of the Interior, Washington, DC.
Morris, Earl
1939 Archaeological Studies in the La Plata District. Carnegie Institution of Washington, Pub. 519. Washington, DC.
Mrozowski, Stephen A.
2010 New and Forgotten Paradigms: The Environment and Economics in Historical Archaeology. *Historical Archaeology* 44(3):117–127.
Mrozowski, Stephen A., Grace H. Ziesing, and Mary C. Beaudry
1996 *Living on the Boott: Historical Archaeology at the Boott Mills Boardinghouses, Lowell, Massachusetts*. University of Massachusetts Press, Amherst.
Muller, Jon D.
1978 The Southeast. In *Ancient Native Americans*, edited by Jesse D. Jennings, pp. 281–327. W. H. Freeman, San Francisco.
1997 *Mississippian Political Economy*. Plenum Press, New York.
Munson-Scullin, Wendy, and Michael Scullin
2005 Potential Productivity of Midwestern Native American Gardens. *Plains Anthropologist* 50:9–21.

Nabokov, Peter

2002 *A Forest of Time: American Indian Ways of History*. Cambridge University Press, New York.

Naroll, Raoul

1962 Floor Area and Settlement Population. *American Antiquity* 27(4):587–589.

Nassaney, Michael S.

1994 Historical and Archaeological Context of Plum Bayou Culture in Central Arkansas. *Southeastern Archaeology* 13(1):36–55.

2003 *Report on the 2003 Archaeological Investigations of the Pointe aux Barques Lighthouse Complex (20HU203)*. Manuscript on file with the Pointe aux Barques Lighthouse Society, Port Hope, Michigan.

National Register of Historic Places

1986 Sampson County Multiple Resource Nomination, NRHP No. 64000465, continuation sheet D, item no. 8, p. 29.

Natural Resources Conservation Service, US Department of Agriculture

2013 Official Soil Series Descriptions. Electronic document, https://www.nrcs.usda.gov, accessed October 10, 2013.

Neitzel, Jill

2003 Three Questions about Pueblo Bonito. In *Pueblo Bonito: Center of the Chacoan World*, edited by Jill Neitzel, pp. 1–9. Smithsonian Institution Press, Washington, DC.

Neubauer, Wolfgang

2004 GIS in Archaeology—the Interface between Prospection and Excavation. *Archaeological Prospection* 11:159–166.

Newman, Christine

1997 *Identification and Discussion of Artifacts from Letchworth Mounds (8JE337)*. Document on file at the Florida Master Site File, Tallahassee.

Okamura, Katsuyuki, and Akira Matsuda (editors)

2011 *New Perspectives in Global Public Archaeology*. Springer, New York.

Orr, Kenneth G.

1946 The Archaeological Situation at Spiro, Oklahoma; a Preliminary Report. *American Antiquity* 11:228–256.

Ortman, Scott G., Shanna R. Diederichs, and Kristin A. Kuckelman

2011 *A Proposal to Conduct Archaeological Testing at Indian Camp Ranch, Montezuma County, Colorado*. Proposal submitted to the Colorado State Historic Preservation Office, Denver. Manuscript on file, Crow Canyon Archaeological Center, Cortez, Colorado.

Pacheco, Paul J., Jarrod D. Burks, and Dee Anne Wymer

2005 *Investigating Ohio Hopewell Settlement Patterns in Central Ohio: A Preliminary Report of Archaeology at Brown's Bottom #1 (33Ro21)*. Ohio Archaeological Council.

2009a *The 2006 Archaeological Investigations at Brown's Bottom #1 (33Ro1104)*. Ohio Archaeological Council.

2009b *The 2007–2008 Archaeological Investigations at Lady's Run (33Ro1105)*. Ohio Archaeological Council.

Palkovich, Ann M.

1975 Point Pattern Analysis and Limited-Activity Sites: An Example from Survey. Master's thesis, Department of Anthropology, Northwestern University, Evanston, Illinois.

Papadopoulos, Nikos G., Apostolos Sarris, William A. Parkinson, Attila Gyucha, Richard W. Yerkes, Paul R. Duffy, and Panagiotis Tsourlos
2014 Electrical Resistivity Tomography for the Modelling of Cultural Deposits and Geomorphological Landscapes at Neolithic Sites: A Case Study from Southeastern Hungary. *Archaeological Prospection* 21:169–183.
Papadopoulos, Nikos G., and Panagiotis Tsourlos
2007 Efficient ERT Measuring and Inversion Strategies for 3D Imaging of Buried Antiquities. *Near Surface Geophysics* 5:346–362.
Patch, Shawn M., and Danny Gregory
2009 *Geophysical Survey of Site 40Mi70 for the CB&I Nuclear Fabrication Facility, Marion County, Tennessee.* New South Associates, Inc., Stone Mountain, Georgia.
2010 Refining Middle Woodland Site Structure with Ground Penetrating Radar: A Case Study from Site 40Mi70, Marion County, Tennessee. Paper Presented at the 75th Annual Meeting of the Society for American Archaeology, Saint Louis, Missouri.
Patch, Shawn M., and Sarah Lowry
2013 *Investigating Prehistoric Settlement in the Upper Guntersville Basin: Archaeological Geophysics Survey of the Raccoon Creek Wildlife Management Area, Jackson County, Alabama.* New South Associates, Inc., Stone Mountain, Georgia.
2014 *Archaeological Geophysics Survey of the Bell Site (40RE1), Roane County, Tennessee.* New South Associates, Inc., Stone Mountain, Georgia.
Patch, Shawn M., Sarah Lowry, and Erin Pritchard
2014 Mississippian Feature Patterning, Site Structure, and Community Organization at the Cox Site (1JA176), Jackson County, Alabama: Geophysical Survey Results. *Early Georgia* 42(2):211–237.
Patch, Shawn M., Sarah Lowry, and Lynne P. Sullivan
2015 *Geophysical Survey of Hiwassee Island (40MG31), Meigs County, Tennessee.* New South Associates, Inc., Stone Mountain, Georgia.
Pauketat, Timothy R., and Thomas E. Emerson
1997 *Cahokia: Domination and Ideology in the Mississippian World.* University of Nebraska Press, Lincoln.
Perino, Gregory H.
1968 The Pete Klunk Mound Group, Calhoun County, Illinois: The Archaic and Hopewell Occupations. In *Hopewell and Woodland Site Archaeology in Illinois*, edited by James A. Brown, pp. 9–124. Illinois Archaeological Survey, Inc., Urbana.
1973a The Koster Mounds, Greene County, Illinois. In *Late Woodland Site Archaeology in Illinois I*, edited by James A. Brown, pp. 141–210. Illinois Archaeological Survey, Inc., Urbana.
1973b The Late Woodland Component at the Pete Klunk Site, Calhoun County, Illinois. In *Late Woodland Site Archaeology in Illinois I*, edited by James A. Brown, pp. 58–89. Illinois Archaeological Survey, Inc., Urbana.
1973c The Late Woodland Component at the Schild Sites, Greene County, Illinois. In *Late Woodland Site Archaeology in Illinois I*, edited by James A. Brown, pp. 90–140. Illinois Archaeological Survey, Inc., Urbana.
2006 *Illinois Hopewell and Late Woodland Mounds: The Excavations of Gregory Perino, 1950–1975.* Studies in Archaeology No. 4. Illinois Transportation Archaeological Research Program, University of Illinois, Urbana.

Perttula, Timothy K.
1992 *"The Caddo Nation": Archaeological and Ethnohistoric Perspectives*. University of Texas Press, Austin.
2005 1938–1939 WPA Excavations at the Hatchel Site (41BW3) on the Red River in Bowie, County, Texas. *Southeastern Archaeology* 24:180–198.
2009 Extended Entranceway Structures in the Caddo Archaeological Area. *Southeastern Archaeology* 28(1):27–42.
2010 Papers on Geophysical Investigations of Woodland and Caddo Sites in the Caddo Area of the Southeastern United States. *Southeastern Archaeology* 29(2):233–235.
Perttula, Timothy K., Bo Nelson, Mark Walters, Robert Cast, and Bobby Gonzalez
2009 Documentation of Caddo Funerary Objects in the Gilcrease Museum Collections. *Friends of Northeast Texas Archaeology* 12:1–307.
Perttula, Timothy K., and Robert Rogers
2007 The Evolution of a Caddo Community in Northeastern Texas: The Oak Hill Village Site (41RK214), Rusk County, Texas. *American Antiquity* 72(1):71–94.
2012 The Evolution of a Caddo Community in Northeastern Texas. In *The Archaeology of the Caddo*, edited by Timothy K. Perttula and Chester P. Walker, pp. 209–238. University of Nebraska Press, Lincoln.
Perttula, Timothy K., and Bob D. Skiles
2014 The Construction and Eventual Burning of the Experimental Caddo House Structure at the George C. Davis Site (41CE19) in East Texas. *Bulletin of the Texas Archeological Society* 85:33–48.
Perttula, Timothy K., and Chester P. Walker (editors)
2012 *The Archaeology of the Caddo*. University of Nebraska Press, Lincoln.
Perttula, Timothy K., Chester P. Walker, and T. Clay Schultz
2008 A Revolution in Caddo Archaeology: The Remote Sensing and Archaeological View from the Hill Farm Site (41BW169) in Bowie County, Texas. *Southeastern Archaeology* 27(1):93–107.
Peukert, John N.
2002 Ground Penetrating Radar at Hollywood. Unpublished master's thesis, University of Mississippi, Oxford.
Phillips, Phillip
1970 Archaeological Survey in the Lower Yazoo Basin, Mississippi, 1949–1955. *Papers of the Peabody Museum of Archaeology and Ethnology* 60(1–2).
Phillips, Philip, James A. Ford, and James B. Griffin
1951 *Archaeological Survey in the Lower Mississippi Alluvial Valley, 1940–1947*. Peabody Museum Papers Vol. 25. Harvard University, Cambridge, Massachusetts.
Plog, Stephen
2008 *Ancient Peoples of the American Southwest*. Thames and Hudson, London.
Pluckhahn, Thomas J.
2003 *Kolomoki: Settlement, Ceremony, and Status in the Deep South, A.D. 350 to 750*. University of Alabama Press, Tuscaloosa.
2010 Household Archaeology in the Southeastern United States: History, Trends, and Challenges. *Journal of Archaeological Research* 18:331–385.
Pluckhahn, Thomas J., and Victor D. Thompson
2009 Mapping Crystal River: Past, Present, Future. *Florida Anthropologist* 62:3–22.
2012 Integrating LiDAR and Conventional Mapping of the Fort Center Site in South-Central Florida: A Comparative Approach. *Journal of Field Archaeology* 37:289–301.

2014 Monumentality beyond Scale: The Elaboration of Mounded Architecture at Crystal River. In *Precolumbian Archaeology in Florida: New Approaches to the Appendicular Southeast*, edited by Neill Wallis and Asa Randall, pp. 62–73. University Press of Florida, Gainesville.

Pluckhahn, Thomas J., Victor D. Thompson, and Brent R. Weisman

2010 Toward a New View of History and Process at Crystal River (8CH11). *Southeastern Archaeology* 29(1):164–181.

Polhemus, Richard

1987 *The Toqua Site–40MR6: A Late Mississippian, Dallas Phase Town*. University of Tennessee Department of Anthropology, Report of Investigations No. 41 and Tennessee Valley Authority Publications in Anthropology No. 44, Knoxville.

Pollack, David

2004 *Caborn-Welborn: Constructing a New Society after the Angel Chiefdom Collapse*. University of Alabama Press, Tuscaloosa.

Powell, J. W.

1884 *Third Annual Report of the Bureau of American Ethnology*. Smithsonian Institution, Washington, DC.

Prentiss, Anna Marie, Guy Cross, Thomas A. Foor, Mathew Hogan, Dirk Markle, and David S. Clarke

2008 Evolution of a Late Prehistoric Winter Village on the Interior Plateau of British Columbia: Geophysical Investigations, Radiocarbon Dating, and Spatial Analysis of the Bridge River Site. *American Antiquity* 73(1):59–82.

Prentiss, Anna Marie, Thomas A. Foor, Guy Cross, Lucille E. Harris, and Michael Wanzenried

2012 The Cultural Evolution of Material Wealth-Based Inequality at Bridge River, British Columbia. *American Antiquity* 77(3):542–564.

Price, Douglas T., and Anne B. Gebauer (editors)

1995 New Perspectives on the Transition to Agriculture. In *Last Hunters First Farmers*. School of American Research, Santa Fe, New Mexico.

Pritchard, Erin

2009 *TVA Archaeology: Seventy-Five Years of Prehistoric Site Research*. University of Tennessee Press, Knoxville.

Pulsepower Developments

2009 *Operating Instructions for the MVM Magnetic Viscosity Meter*. Pulsepower Developments, Oxford, United Kingdom.

QGIS Development Team

2009 *QGIS Geographic Information System*. Open Source Geospatial Foundation.

Rafferty, Janet R., and Evan Peacock (editors)

2008 *Time's River: Archaeological Syntheses from the Lower Mississippi River Valley*. University of Alabama Press, Tuscaloosa.

Rapoport, Amos

1982 *The Meaning of the Built Environment*. University of Arizona Press, Tucson.

Rees, Mark A., and Patrick C. Livingood (editors)

2007 *Plaquemine Archaeology*. University of Alabama Press, Tuscaloosa.

Regnier, Amanda L., Scott W. Hammerstedt, and Nicholas H. Beale

2014 The Grobin Davis Site: Archaeogeophysics and Settlement Patterns at Caddo Mound Centers in Southeastern Oklahoma. *Southeastern Archaeology* 33:87–107.

Reilly, Kent F., III, and James F. Garber (editors)
2007 *Ancient Objects and Sacred Realms: Interpretations of Mississippian Iconography*. University of Texas Press, Austin.
Relph, Edward
1976 *Place and Placelessness*. Pion, Ltd., Thousand Oaks, California.
Reynolds, Matthew D.
2002 Magnetic Remote Sensing and Ground Truth: Some Examples from the Hollywood Site, Tunica County, Mississippi. Unpublished master's thesis, University of Mississippi, Oxford.
Robb, John G.
1998 The "Ritual Landscape" Concept in Archaeology: A Heritage Construction. *Landscape Research* 23(2):159–175.
Roberts, Ricky L.
1977 Population Estimates. In *The Talking Crow Site*, edited by Carlyle S. Smith, pp. 167–176. Publications in Anthropology No. 9. University of Kansas, Lawrence.
Robinson, Hugh L.
2014 *Data Recovery at 5MT20065 for Empire Electric Association's Operations Center, Montezuma County, Colorado*. Woods Canyon Archaeological Consultants, Cortez, Colorado.
Robinson, Kenneth W.
1997 Port Brunswick and the Colonial Naval Stores Industry: Historical and Archaeological Observations. *North Carolina Archaeology* 46:51–68.
Rochelo, Mark J., Christian Davenport, and Donna Selch
2015 Revealing Pre-historic Native American Belle Glade Earthworks in the Northern Everglades Utilizing Airborne LiDAR. *Journal of Archaeological Science: Reports* 2:624–643.
Rogers, J. Daniel
1980 *Spiro Archaeology: 1979 Excavations*. Oklahoma Archeological Survey Studies in Oklahoma's Past 6, Norman.
1982a *Spiro Archaeology: 1980 Research*. Oklahoma Archeological Survey Studies in Oklahoma's Past 9, Norman.
1982b *Spiro Archaeology: The Plaza*. Oklahoma Archeological Survey Studies in Oklahoma's Past 10, Norman.
2011 Spiro and the Development of a Regional Social System. In *Artifacts from the Craig Mound at Spiro, Oklahoma*, edited by April K. Sievert and J. Daniel Rogers, pp. 1–12. Smithsonian Institution Contributions to Anthropology 49, Washington, DC.
Rogers, J. Daniel, Don G. Wyckoff, and Dennis A. Peterson (editors)
1989 *Contributions to Spiro Archeology: Mound Excavations and Regional Perspectives*. Oklahoma Archeological Survey Studies in Oklahoma's Past 16, Norman.
Rolingson, Martha Ann
1982 *Emerging Patterns of Plum Bayou Culture: Preliminary Investigations of the Toltec Mounds Research Project*. Arkansas Archeological Survey Research Series No. 18. Arkansas Archeological Survey, Fayetteville.
2002 Plum Bayou Culture of the Arkansas-White River Basin. In *The Woodland Southeast*, edited by David G. Anderson and Robert C. Mainfort Jr., pp. 44–65. University of Alabama Press, Tuscaloosa.
2012 *Toltec Mounds: Archeology of the Mound-and-Plaza Complex*. Arkansas Archeological Survey Research Series No. 65. Arkansas Archeological Survey, Fayetteville.

Roper, Donna C., and Elizabeth P. Pauls

2005 What, Where, and When Is an Earthlodge? In *Plains Earthlodges: Ethnographic and Archaeological Perspectives*, edited by Donna C. Roper and Elizabeth P. Pauls, pp. 1–31. University of Alabama Press, Tuscaloosa.

Rotman, Deborah L.

2003 Introduction: Exploring Shared Spaces and Divided Places on the American Historical Landscape. In *Shared Spaces and Divided Places: Exploring the Material and Spatial Dimensions of Gender Relations and American Historical Landscapes*, edited by Deborah L. Rotman and Ellen R. Savulis, pp. 1–23. University of Tennessee Press, Knoxville.

Rotman, Deborah L., and Michael S. Nassaney

1997 Class, Gender, and the Built Environment: Deriving Social Relations from Cultural Landscapes in Southwest Michigan. *Historical Archaeology* 31(2):42–62.

Rowlands, Aled, and Apostolos Sarris

2007 Detection of Exposed and Subsurface Archaeological Remains Using Multisensor Remote Sensing. *Journal of Archaeological Science* 34:795–803.

Ruby, Bret J., Christopher Carr, and Douglas K. Charles

2005 Community Organizations in the Scioto, Mann, and Havana Hopewellian Regions: A Comparative Perspective. In *Gathering Hopewell: Society, Ritual, and Ritual Interaction*, edited by Christopher Carr and D. Troy Case, pp. 119–176. Kluwer Academic, New York.

Ruhl, Donna L.

2000 Archaeobotany at Bernath Place (8SR986) and Other Santa Rosa/Swift Creek Related Sites in Coastal and Non-coastal Southeastern U.S. Locations. *Florida Anthropologist* 53(2–3):190–202.

Russo, Michael, Craig Dengel, and Jeffrey Shanks

2014 Northwest Florida Woodland Mounds and Middens. In *New Histories of Pre-Columbian Florida*, edited by Neill J. Wallis and Asa R. Randall, pp. 121–142. University Press of Florida, Gainesville.

Russo, Michael, Carla S. Hadden, and Craig Dengel

2009 *Archaeological Investigations of Mounds and Ring Middens at Hark Hammock Tyndall Air Force Base*. Document on file at the Southeast Archeological Center, National Park Service, Tallahassee, Florida.

Sabo, George, III

2008 Rock Art and the Study of Ancient Religions in Southeastern North America. In *Religion, Archaeology, and the Material World*, edited by Lars Fogelin, pp. 279–296. Occasional Paper No. 36. Center for Archaeological Investigations, Southern Illinois University, Carbondale.

2012 The Terán Map and Caddo Cosmology. In *The Archaeology of the Caddo*, edited by Timothy K. Perttula and Chester P. Walker, pp. 431–448. University of Nebraska Press, Lincoln.

Scarry, C. Margaret

1998 Domestic Life on the Northwest Riverbank at Moundville. In *Archaeology of the Moundville Chiefdom*, edited by Vernon J. Knight Jr. and Vincas P. Steponaitis, pp. 63–101. Smithsonian Institution Press, Washington, DC.

Schachner, Gregson, Dennis Gilpin, and Matthew Peoples

2012 Alternative Trajectories during the Early Pueblo Period in the Little Colorado Drainage and Beyond. In *Crucible of the Pueblos*. Cotsen Institute of Archaeology, University of California, Los Angeles.

Schambach, Frank F.

1982 The Archeology of the Great Bend Region in Arkansas. In *Contributions to the Archeology of the Great Bend Region*, edited by Frank F. Schambach and Frank Rackerby, pp. 1–11. Research Series No. 22. Arkansas Archeological Survey, Fayetteville.

1996 Mounds, Embankments, and Ceremonialism in the Trans-Mississippi South. In *Mounds, Embankments, and Ceremonialism in the Midsouth*, edited by Robert C. Mainfort and Richard Walling, pp. 36–43. Research Series No. 46. Arkansas Archeological Survey, Fayetteville.

Schambach, Frank. F., John E. Miller III, and David B. Kelley

1980 Original unpublished field notes from Battle Mound (3LA1). Manuscript on file, Arkansas Archeological Survey, Fayetteville.

Schlanger, Sarah H.

1992 Recognizing Persistent Places in Anasazi Settlement Systems. In *Space, Time, and Archaeological Landscapes*, edited by Jaqueline Rossignol and LuAnn Wandsnider, pp. 91–112. Plenum Press, New York.

Schmidt, Armin

2013 *Earth Resistance for Archaeologists*. AltaMira Press, Lanham, Maryland.

Schnell, Frank T., Vernon J. Knight Jr., and Gail Schnell

1981 *Cemochechobee: Archaeology of a Mississippian Ceremonial Center on the Chattahoochee River*. University Press of Florida, Tallahassee.

Schroeder, Sissel

2011 An Investigation of the Origins of Variation in Perishable Architecture at Jonathan Creek. *Southeastern Archaeology* 30(2):311–336.

2013 Culture, Time, and Practice: The Shifting Interpretive Potential of New Deal-Era Collections. In *Shovel Ready: Archaeology and Roosevelt's New Deal for America*, edited by Bernard K. Means, pp. 165–184. University of Alabama Press, Tuscaloosa.

Schroedl, Gerald F.

1998 Mississippian Towns in the Eastern Tennessee Valley. In *Mississippian Towns and Sacred Spaces: Searching for an Architectural Grammar*, edited by R. Barry Lewis and Charles Stout, pp. 138–143. University of Alabama Press, Tuscaloosa.

2009 The Tellico Archaeological Project. In *TVA Archaeology: Seventy-Five Years of Prehistoric Site Research*, edited by Erin Pritchard, pp. 63–110. University of Tennessee Press, Knoxville.

Schroedl, Gerald F., C. Clifford Boyd, and R. P. Stephen Davis

1990 Explaining Mississippian Origins in East Tennessee. In *The Mississippian Emergence*, edited by Bruce D. Smith, pp. 175–196. Smithsonian Institution Press, Washington, DC.

Schroedl, Gerald F., R. P. Stephen Davis, and C. Clifford Boyd

1985 *Archaeological Contexts and Assemblages at Martin Farm*. University of Tennessee, Knoxville.

Schultz, T. Clay

2010 Architectural Variability in the Caddo Area of Eastern Texas. *Friends of Northeast Texas Archaeology* 16:1–439.

Schwadron, Margo

2010a Landscapes of Maritime Complexity: Prehistoric Shell Work Sites of the Ten Thousand Islands, Florida. Unpublished PhD dissertation, Department of Archaeology and Ancient History, University of Leicester.

2010b Prehistoric Landscapes of Complexity: Archaic and Woodland Period Shell

Works, Shell Rings, and Tree Islands of the Everglades, South Florida. In *Trend, Tradition, and Turmoil: What Happened to the Southeastern Archaic*, edited by David Hurst Thomas and Matthew C. Sanger, pp. 113–147. Anthropological Papers of the American Museum of Natural History Vol. 93, New York.

Schwartz, Douglas W.

1980 *Archaeology of the Grand Canyon: Unkar Delta*. Grand Canyon Archaeological Series, Volume 2. School of American Research Press, Santa Fe, New Mexico.

2008 Archaeology of the Grand Canyon: A Personal Look Back. In *Reflections of Grand Canyon Historians: Ideas, Arguments, and First-Person Accounts*, edited by Todd R. Berger, pp. 199–204. Grand Canyon Association, Grand Canyon, Arizona.

Scollar, Irwin, A. Tabbagh, A. Hesse, and I. Herzog

1990 *Archaeological Prospecting and Remote Sensing*. Cambridge University Press, Cambridge.

Sears, William H.

1958 The Wilbanks Site (9CK-5), Georgia. *Bureau of American Ethnology Bulletin* 169: 129–194. Washington, DC.

Sebastian, Lynne

1992 Chaco Canyon and the Anasazi Southwest: Changing Views of Sociopolitical Organization. In *Anasazi Regional Organization and the Chaco System*, edited by David E. Doyel, pp. 23–32. Maxwell Museum of Anthropology Anthropological Papers No. 5. Albuquerque, New Mexico.

Seinfeld, Daniel M., and Daniel P. Bigman

2013 *Investigating Monumental Architecture at the Letchworth Mounds Site (8JE337)*. Manuscript on file, Florida Department of State, Division of Historical Resources, Tallahassee.

Sherwood, Sarah C., and T. R. Kidder

2011 The DaVincis of Dirt: Geoarchaeological Perspectives on Native American Mound Building in the Mississippi River Basin. *Journal of Anthropological Archaeology* 30: 69–87.

Sievert, April K., and J. Daniel Rogers

2011 *Artifacts from the Craig Mound at Spiro, Oklahoma*. Smithsonian Institution Scholarly Press, Washington, DC.

Simmons, Alan H.

2007 *The Neolithic Revolution in the Near East: Transforming the Human Landscape*. University of Arizona Press, Tucson.

Smith, Bruce D.

1978 Variations in Mississippian Settlement Patterns. In *Mississippian Settlement Patterns*, edited by Bruce D. Smith, pp. 478–503. Academic Press, New York.

1989 Origins of Agriculture in Eastern North America. *Science* 246(4937):1566–1571.

Smith, Bruce D. (editor)

1990 *The Mississippian Emergence*. Smithsonian Institution Press, Washington, DC.

Smith, Michael E.

2007 Form and Meaning in the Earliest Cities: A New Approach to Ancient Urban Planning. *Journal of Planning History* 6:3–47.

Snyder, John Francis

1895 Prehistoric Illinois: Certain Indian Mounds Technically Considered. *Journal of the Illinois State Historical Society* 1(4) and 2(1–2):31–40, 47–69, 72–92.

1909 A Group of Illinois Mounds. *The Archaeologist* 3(4):109–113.

Sofaer, Anna
2007 The Primary Architecture of the Chacoan Culture: A Cosmological Expression. In *Architecture of Chaco Canyon, New Mexico*, edited by Stephen A. Lekson, pp. 225–254. University of Utah Press, Salt Lake City.
Somers, Lewis
2006 Resistivity Survey. In *Remote Sensing in Archaeology: An Explicitly North American Perspective*, edited by Jay K. Johnson, pp. 109–129. University of Alabama Press, Tuscaloosa.
Southerling, Bobby
2013 *A Survey of Tar and Pitch Production Sites in Francis Marion National Forest*. Paper presented at the 2013 meeting Archaeology in North Carolina: A Retrospective of Recent Research, Chapel Hill.
Spoerry, Paul (editor)
1992 *Geoprospection in the Archaeological Landscape*. Oxbow Monograph No. 18. Oxbow Books, Oxford, United Kingdom.
Stallings, Richard
1994 A Final Report of Investigations at the Hollywood Site (22TU500). Paper presented at the 1994 Southeastern Archaeological Conference, Lexington, Kentucky.
Steere, Benjamin A.
2011 The Archaeology of Houses and Households in the Native Southeast. Unpublished PhD dissertation, University of Georgia, Athens.
Stein, John R., and Andrew P. Fowler
1996 Looking beyond Chaco in the San Juan Basin and Its Peripheries. In *The Prehistoric Pueblo World A.D. 1150–1350*, edited by Michael A. Adler, pp. 114–130. University of Arizona Press, Tucson.
Stein, John R., Richard Friedman, Taft Blackhorse, and Richard W. Loose
2007 Revisiting Downtown Chaco. In *The Architecture of Chaco Canyon, New Mexico*, edited by Stephen H. Lekson, pp. 199–223. University of Utah Press, Salt Lake City.
Steponaitis, Vincas P.
1986 Prehistoric Archaeology in the Southeastern United States, 1970–1985. *Annual Review of Anthropology* 15:363–404.
2009 *Ceramics, Chronology, and Community Patterns: An Archaeological Study at Moundville*. 2nd ed. University of Alabama Press, Tuscaloosa.
Sternberg, Robert S.
1987 Archaeomagnetism and Magnetic Anomalies in the American Southwest. *Geophysics* 52(3):368–371.
Stewart, Ian
2013 *Symmetry: A Very Short Introduction*. Oxford University Press, New York.
Stottman, M. Jay (editor)
2010 *Archaeologists as Activists: Can Archaeologists Change the World?* University of Alabama Press, Tuscaloosa.
Struever, Stewart
1960 The Kamp Mound Group and a Hopewell Mortuary Complex in the Lower Illinois Valley. Master's thesis, Department of Anthropology, Northwestern University, Evanston, Illinois.
1964 The Hopewell Interaction Sphere in Riverine-Western Great Lakes Culture History. In *Hopewellian Studies*, vol. 12(3), edited by Joseph R. Caldwell and Robert L. Hall, pp. 81–86. Illinois State Museum Scientific Papers.

Struever, Stewart, and Gail L. Houart
1972 An Analysis of the Hopewell Interaction Sphere. In *Social Exchange and Interaction*, edited by Edwin N. Wilmsen, pp. 47–79. Museum of Anthropology, University of Michigan, Ann Arbor.
Sturm, Jennie O., and Patricia L. Crown
2015 Micro-scale Mapping Using Ground-Penetrating Radar: An Example from Room 28, Pueblo Bonito, Chaco Canyon, New Mexico. *Advances in Archaeological Practice* 3(2):124–135.
Sullivan, Alan P., III, Kevin S. Magee, Philip B. Mink II, and Kathleen M. Forste
2012 Remote Sensing of Heritage Resources for Research and Management. *Park Science* (Winter 2011/2012):93–98.
Sullivan, Alan P., III, Philip B. Mink II, and Patrick M. Uphus
2002 From John W. Powell to Robert C. Euler: Testing Models of Grand Canyon's Prehistoric Puebloan Settlement History. In *Culture and Environment in the American Southwest*. SWCA Anthropological Research Paper No. 8. SWCA Environmental Consultants, Phoenix.
Sullivan, Lynne P.
1987 The Mouse Creek Phase Household. *Southeastern Archaeology* 6:16–29.
1995 Mississippian Household and Community Organization in Eastern Tennessee. In *Mississippian Communities and Households*, edited by J. Daniel Rogers and Bruce D. Smith, pp. 99–123. University of Alabama Press, Tuscaloosa.
2007 Shell Gorgets, Time, and the SECC in Southeastern Tennessee. In *Southeastern Ceremonial Complex: Chronology, Content, Context*, edited by Adam King, pp. 88–106. University of Alabama Press, Tuscaloosa.
2009 Archaeological Time Constructs and the Construction of the Hiwassee Island Mound. In *TVA Archaeology: Seventy-Five Years of Prehistoric Site Research*, edited by Erin Pritchard, pp. 181–209. University of Tennessee Press, Knoxville.
2016 Reconfiguring the Chickamauga Basin. In *New Deal Archaeology in Tennessee*, edited by David Dye, pp. 138–170. University of Alabama Press, Tuscaloosa.
Sullivan, Stephanie M., and Duncan P. McKinnon
2013 The Collins Site (3WA1): Exploring Architectural Variation in the Western Ozark Highlands. *Southeastern Archaeology* 32:70–84.
Taylor, Jay L. B.
1928 Archaeological Exploration of the Lower Illinois Valley (March 26 to October 1, 1928). In *Early Hopewell Mound Explorations: The First Fifty Years in the Illinois River Valley*, edited by Kenneth B. Farnsworth, pp. 271–316. Illinois Transportation Archaeological Research Program, University of Illinois, Urbana.
Tesar, Louis D., Brenda N. Swann, James J. Miller, Melanie Damour-Horrel, and Michael Lavender
2003 *Results of the Letchworth Mounds (8JE337) Archaeological State Park Auger and Topographic Survey with Management Recommendations*. Document on file at the Florida Master Site File, Tallahassee.
Thoburn, Joseph B.
1931 Prehistoric Cultures of Oklahoma. In *Archaeology of the Arkansas River Valley*, edited by Warren K. Moorehead, pp. 53–130. Yale University Press, New Haven, Connecticut.
Thomas, Cyrus
1887 *Burial Mounds of the Northern Sections of the United States*. Bulletin No. 4, Bureau of American Ethnology. Smithsonian Institution Press, Washington, DC.

1894 *Report on the Mound Exploration of the Bureau of Ethnology, 12th Annual Report of the Bureau of Ethnology to the Secretary of the Smithsonian Institution, 1890–91*. Smithsonian Institution Press, Washington, DC.

Thompson, Raymond H.
2004 Glimpses of the Young Emil Haury. *Journal of the Southwest* 46(1):3–7.

Thompson, Victor D.
2014 Monumental Architecture, Households, and Archaeo-Geophysics in the American Southeast and Midwest. *Archaeological Prospection* 21:1–3.

Thompson, Victor D., and C. Fred T. Andrus
2011 Evaluating Mobility, Monumentality, and Feasting at the Sapelo Island Shell Ring Complex. *American Antiquity* 76:315–344.

Thompson, Victor D., Philip J. Arnold III, Thomas J. Pluckhahn, and Amber M. VanDerwarker
2011 Situating Remote Sensing in Anthropological Archaeology. *Archaeological Prospection* 18:195–213.

Thompson, Victor D., Kristen Gremillion, and Thomas J. Pluckhahn
2013 Challenging the Evidence for Prehistoric Wetland Maize Agriculture at Fort Center, Florida. *American Antiquity* 78:181–193.

Thompson, Victor D., and William Marquardt
2015 *Report on Archaeological Investigations at Mound Key (8LL2) (2013 and 2014 Field Seasons), Lee County, Florida*. Committee for Research and Exploration, National Geographic Society, Washington, DC.

Thompson, Victor D., William H. Marquardt, and Karen J. Walker
2014 A Remote Sensing Perspective on Shoreline Modification, Canal Construction, and Household Trajectories at Pineland along Florida's Southwestern Gulf Coast. *Archaeological Prospection* 21:59–73.

Thompson, Victor D., and Thomas J. Pluckhahn
2012 Monumentalization and Ritual Landscapes at Fort Center in the Lake Okeechobee Basin of South Florida. *Journal of Anthropological Archaeology* 31:49–65.
2014 The Modification and Manipulation of Landscape at Fort Center. In *Precolumbian Archaeology in Florida: New Approaches to the Appendicular Southeast*, edited by Neill Wallis and Asa Randall, pp. 163–182. University Press of Florida, Gainesville.

Thompson, Victor D., and John Worth
2011 Dwellers by the Sea: Native American Coastal Adaptations along the Southern Coasts of Eastern North America. *Journal of Archaeological Research* 19(1):51–101.

Tite, M. S., and C. E. Mullins
1971 Enhancement of the Magnetic Susceptibility of Soils on Archaeological Sites. *Archaeometry* 12:209–219.

Toom, Dennis L., and Kenneth L. Kvamme
2002 The "Big House" at Whistling Elk Village (39HU242): Geophysical Findings and Archaeological Truths. *Plains Anthropologist* 47(180):5–16.

Torrence, Corbett M., Samuel J. Chapman, and William H. Marquardt
1994 *Topographic Mapping and Archaeological Reconnaissance of Mound Key State Archaeological Site (8LL2), Estero Bay, Florida*. Institute of Archaeology and Paleoenvironmental Studies, Florida Museum of Natural History, University of Florida, Gainesville.

Trigger, Bruce G.
2006 *A History of Archaeological Thought*. Cambridge University Press, New York.

TR Systems, Ltd.
2002 *TR/CIA Resistance Meter Instruction Manual.* TR Systems, Ltd., Peterborough, United Kingdom.
Trubitt, Mary Beth
2009 Burning and Burying Buildings: Exploring Variation in Caddo Architecture in Southwest Arkansas. *Southeastern Archaeology* 28(2):233–247.
Trubowitz, Neal L. (editor)
1984 *Cedar Grove.* Arkansas Archeological Survey Research Series No. 23. Arkansas Archeological Survey, Fayetteville.
Tsourlos, Panagiotis
1995 *Modelling, Interpretation and Inversion of Multielectrode Resistivity Survey Data.* University of York, United Kingdom.
Tuan, Yi-Fu
1977 *Space and Place: The Perspective of Experience.* University of Minnesota Press, Minneapolis.
Turck, John A.
2013 *Protocols for Downloading and Processing LiDAR Data.* Manuscript on file at the University of Georgia, Laboratory of Archaeology, Athens.
Valastro, S., Jr., E. Mott Davis, and Alejandra G. Varela
1972 University of Texas at Austin Radiocarbon Dates IX. *Radiocarbon* 14(2):461–485.
Van Dyke, Ruth M.
2007 *The Chaco Experience: Landscape and Ideology at the Center Place.* School for Advanced Research Press, Santa Fe, New Mexico.
Van Gilder, Cynthia L., and Douglas K. Charles
2003 Archaeology as Cultural Encounter: The Legacy of Hopewell. In *Theory, Method, and Practice in Modern Archaeology*, edited by Robert J. Jeske and Douglas K. Charles, pp. 114–129. Praeger, Westport, Connecticut.
Van Nest, Julieann
2006 Rediscovering This Earth: Some Ethnogeological Aspects of the Illinois Valley Hopewell Mounds. In *Recreating Hopewell*, edited by Douglas K. Charles and Jane E. Buikstra, pp. 402–426. University Press of Florida, Gainesville.
Van Nest, Julieann, and David L. Asch
2001 The Illinois Hopewell Mounds at Kamp: Re-interpreting a 40-Year-Old Excavation from a Geoarchaeological Perspective. Paper presented at the 66th Annual Meeting of the Society for American Archaeology, New Orleans, Louisiana.
Van Nest, Julieann, Douglas K. Charles, Jane E. Buikstra, and David L. Asch
2001 Sod Blocks in Illinois Hopewell Mounds. *American Antiquity* 66(4):633–650.
Van West, Carla R.
2011 Landscape Change: Archaeological Perspectives of the Legacy of Human-Environmental Interactions in the U.S. Southwest. In *Movement, Connectivity, and Landscape Change in the Ancient Southwest*, edited by Margaret C. Nelson and Colleen A. Strawhacker, pp. 107–118. University Press of Colorado, Boulder.
Varien, Mark D., and James M. Potter
2008 *The Social Construction of Communities: Agency, Structure, and Identity in Prehispanic Southwest.* AltaMira Press, Lanham, Maryland.
Vickers, Roger S., and Lambert T. Dolphin
1975 A Communication on an Archaeological Radar Experiment at Chaco Canyon, New Mexico. *Museum of Applied Science Center for Archaeology Newsletter* Vol. 2. University of Pennsylvania, Philadelphia.

Vickers, Roger S., Lambert T. Dolphin, and David Johnson
1976 Archeological Investigations at Chaco Canyon Using a Subsurface Radar. In *Remote Sensing Experiments in Cultural Resource Studies: Non-destructive Methods of Archeological Exploration, Survey, and Analysis*. Reports of the Chaco Center 1:81–101. National Park Service and University of New Mexico, Albuquerque.
Vivian, R. Gwinn
1970 An Inquiry into Prehistoric Social Organization in Chaco Canyon, New Mexico. In *Reconstructing Prehistoric Pueblo Societies*, edited by William A. Longacre, pp. 59–83. University of New Mexico Press, Albuquerque.
1974 Conservation and Diversion: Water-Control Systems in the Anasazi Southwest. In *Irrigation's Impact on Society*, edited by Theodore E. Downing and McGuire Gibson, pp. 95–112. Anthropological Papers of the University of Arizona No. 25. University of Arizona Press, Tucson.
1990 *The Chacoan Prehistory of the San Juan Basin*. Academic Press, San Diego.
Vivian, R. Gwinn, Carla R. Van West, Jeffery S. Dean, Nancy J. Akins, Mollie S. Toll, and Thomas C. Windes
2006 Ecology and Economy. In *The Archaeology of Chaco Canyon: An Eleventh-Century Pueblo Regional Center*, edited by Stephen H. Lekson, pp. 45–66. School of American Research Press, Santa Fe, New Mexico.
Vogel, Gregory
2012 Viewshed Characteristics of Caddo Mounds in the Arkansas Basin. In *The Archaeology of the Caddo*, edited by Timothy K. Perttula and Chester P. Walker, pp. 139–176. University of Nebraska Press, Lincoln.
Volf, William
2011 *Letter Report on the Electrical Resistivity and Magnetic Gradient Survey of Eight Sites in Indian Camp Ranch, Montezuma County, Colorado*. Natural Resources Conservation Service. Report on file, Crow Canyon Archaeological Center, Cortez, Colorado.
Walker, Chester P.
2009 Landscape Archaeogeophysics: A Study of Magnetometer Surveys from Etowah (9BR1), the George C. Davis Site (41CE19), and the Hill Farm Site (41BW169). PhD dissertation, Department of Anthropology, University of Texas at Austin.
Walker, Chester P., and Duncan P. McKinnon
2012 Exploring Prehistoric Caddo Communities through Archaeogeophysics. In *The Archaeology of the Caddo*, edited by Timothy K. Perttula and Chester P. Walker, pp. 177–208. University of Nebraska Press, Lincoln.
Walker, Chester, P., and Timothy K. Perttula
2007 Remote Sensing at the Horace Cabe Site (41BW14). *Caddo Archeology Journal* 16:37–44.
2008 Geophysical Investigations on Caddo Sites in East Texas and Surrounding States. *Bulletin of the Texas Archeological Society* 79:159–176.
Wallis, Neill
2011 *The Swift Creek Gift: Vessel Exchange on the Atlantic Coast*. University of Alabama Press, Tuscaloosa.
2013 Swift Creek and Weeden Island Mortuary Landscapes of Interaction. In *Early and Middle Woodland Landscapes of the Southeast*, pp. 204–218. University Press of Florida, Gainesville.
Watters, Margaret
2012 *Remote Sensing Survey Preliminary Report, Dillard Archaeological Site, Crow Canyon, CO*. Oregon Public Broadcasting, Portland.

Watters, Margaret, and Shanna R. Diederichs
2013 Archaeological Geophysical Surveys Reveal the Basketmaker III Population at Dillard Site. *ISAP News: The Newsletter of the International Society for Archaeological Prospection* 35:12–14.
Webb, Clarence H.
1940 House Types among the Caddo Indians. *Bulletin of the Texas Archeological and Paleontological Society* 12:49–75.
1959 The Belcher Mound: A Stratified Caddoan Site in Caddo Parish, Louisiana. *Society for American Archaeology* 16.
1983 The Bossier Focus Revisited: Montgomery I, Werner and Other Unicomponent Sites. In *Southeastern Natives and Their Pasts; Papers Honoring Dr. Robert E. Bell*, edited by Don G. Wyckoff and J. L. Hoffman, pp. 183–240. Oklahoma Archeological Survey Studies in Oklahoma's Past 11, Norman.
Webb, William S.
1938 *An Archaeological Survey of the Norris Basin in Eastern Tennessee*. Bureau of American Ethnology Bulletin No. 118. Smithsonian Institution Press, Washington, DC.
1939 *An Archaeological Survey of the Wheeler Basin on the Tennessee River in Northern Alabama*. US Government Printing Office, Washington, DC.
1952 *The Jonathan Creek Village, Site 4, Marshall County, Kentucky*. Reports in Anthropology 8. Department of Anthropology, University Press of Kentucky, Lexington.
Webb, William S., and David L. DeJarnette
1942 *An Archaeological Survey of the Pickwick Basin in the Adjacent States of Alabama, Mississippi and Tennessee*. Bureau of American Ethnology Bulletin 129. Smithsonian Institution Press, Washington, DC.
Webb, William S., and Charles G. Wilder
1951 *An Archaeological Survey of Guntersville Basin on the Tennessee River in Northern Alabama*. University Press of Kentucky, Lexington.
Wedel, Waldo R.
1979 House Floors and Native Settlement Populations in the Central Plains. *Plains Anthropologist* 24:85–98.
Weymouth, John W.
1986 Geophysical Methods of Archaeological Site Surveying. *Advances in Archaeological Method and Theory* 9:311–395.
Wheeler, Ryan J., and James S. Dunbar
2005 *Archaeology of the Proposed Letchworth Mounds Archaeological State Park Addition and Conservation Easement (Booth Parcel)*. Document on file at the Florida Master Site File, Tallahassee.
Widmer, Randolph. J.
1988 *The Evolution of Calusa: A Nonagricultural Chiefdom of the Southwest Florida Coast*. University of Alabama Press, Tuscaloosa.
1996 Recent Excavations at the Key Marco Site, 8Cr48, Collier County, Florida. *Florida Anthropologist* 49:10–25.
Wiewel, Adam S., Duncan P. McKinnon, and Kenneth L. Kvamme
2009 *Topographic Mapping and Creation of a Digital Elevation Model for the Lynch Knife River Flint Quarry (32DU526), North Dakota*. Report submitted to the State Historical Society of North Dakota.
Wilk, Richard R., and William L. Rathje
1982 Household Archaeology. *American Behavioral Scientist* 25:617–639.

Will, George F., and Thad C. Hecker
1944 The Upper Missouri Valley Aboriginal Culture in North Dakota. *North Dakota Historical Quarterly* 11(1–2):5–126.
Will, George F., and George E. Hyde
1917 *Corn among the Indians of the Upper Missouri.* William Harvey Miner, Saint Louis, Missouri.
Willey, Gordon R.
1949 *Archeology of the Florida Gulf Coast.* AMS Press, Brooklyn, New York.
Willey, Gordon R., and Jeremy A. Sabloff
1993 *A History of American Archaeology.* 3rd ed. Thames and Hudson, London.
Williams, J. Mark, and Daniel T. Elliot (editors).
1998 *A World Engraved: Archaeology of the Swift Creek Culture.* University of Alabama, Tuscaloosa.
Williams, Justin
1935 English Mercantilism and Carolina Naval Stores, 1705–1776. *Journal of Southern History* 1(2):169–185.
Wills, Wirt H., Thomas A. Baker, and Lee A. Baker
1990 Aerial Perspectives on Prehistoric Agricultural Fields of the Middle Rio Grande Valley, New Mexico. In *Clues to the Past: Papers in Honor of William M. Sundt*, edited by Meliha S. Duran and David T. Kirkpatrick. Archaeological Society of New Mexico 16, Albuquerque, New Mexico.
Wilshusen, Richard H.
1999 Basketmaker III (A.D. 500–750). In *Colorado Prehistory: A Context for Southern Colorado Drainage Basin*, edited by William D. Lipe, Mark D. Varien, and Richard H. Wilshusen, pp. 166–195. Colorado Council of Professional Archaeologists, Denver.
Wilson, Gilbert L.
1917 *Agriculture of the Hidatsa Indians: An Indian Interpretation.* Bulletin of the University of Minnesota, Studies in the Social Sciences No. 9. University of Minnesota, Minneapolis.
Wilson, Gregory D.
2008 *The Archaeology of Everyday Life at Early Moundville.* University of Alabama Press, Tuscaloosa.
Wood, W. Raymond
1967 *An Interpretation of Mandan Culture History.* River Basin Surveys Papers No. 39. Smithsonian Institution, Bureau of American Ethnology Bulletin No. 198. US Government Printing Office, Washington, DC.
Wright, Alice P.
2014 History, Monumentality, and Interaction in the Appalachian Summit Middle Woodland. *American Antiquity* 79(2):277–294.
2015 Private Property, Public Archaeology: Resident Communities as Stakeholders in American Archaeology. *World Archaeology* 47(2):212–224.
Wyckoff, Don G.
1970 *A Preliminary Report on 1970 Field Research at the Spiro Site, LeFlore County, Oklahoma.* Oklahoma Archeological Survey.
Yon, J. William, Jr.
1966 Geology of Jefferson County, Florida. *Geological Bulletin* 48. Florida Geological Survey.
Zakšek, Klemen, Krištof Oštir, and Žiga Kokalj
2011 Sky-View Factor as a Relief Visualization Technique. *Remote Sensing* 3:398–415.

# Contributors

*Rory Becker* is an assistant professor of anthropology at Eastern Oregon University in La Grande. At Eastern Oregon University, he established the Cultural Heritage and Archaeological Research Technologies lab (CHART), which provides students with research and applied anthropology opportunities. Becker also has eight years of experience working in the cultural resource management industry for universities, private companies, and state and federal agencies. His research interests currently focus on the use of remote sensing techniques in archaeology and historical archaeology sites, especially those dating to the fur trade and Contact period.

*Daniel P. Bigman* is a visiting lecturer at Georgia State University. He has surveyed numerous sites throughout the southeastern United States using a variety of geophysical survey techniques. He has worked closely with academic institutions and federal and state governments in large-scale mapping projects and cultural protection. His research on geophysical prospection in the Southeast has recently been published in regional and international journals such as *Southeastern Archaeology*, *Archaeological Prospection*, and *Journal of Archaeological Science*.

*Jane E. Buikstra* is Regent's Professor of Bioarchaeology and founding director of the Center for Bioarchaeological Research at Arizona State University, Tempe. She is the chair of the board for the Center for American Archeology. Her areas of specialization include archaeology, forensic anthropology, prehistory of the Americas, skeletal biology, paleopathology, paleodemography, social dimensions of mortuary practice, and biological distance.

*Shanna Diederichs* is a supervisory archaeologist at Crow Canyon Archaeological Center in southwest Colorado and oversees the Basketmaker Communities Project. Her ongoing research focuses on social organization in early agricultural societies in the American Southwest. With a broad background in cultural resource management, the National Park Service, and nonprofit education, she aims to help the public imagine prehistoric societies using geophysical technologies, excavation, and native perspectives.

*Bryan S. Haley* is an archaeologist and terrestrial/marine remote sensing specialist in the New Orleans office of Coastal Environments, Inc. He specializes in prehistoric and historic native archaeology in the southeastern United States. His 16 years of remote sensing experience include projects in 23 American states, Central America, South America, and Europe. Haley has also served as an instructor in a number of remote sensing training seminars, contributed to a variety of publications and reports, and participated in the PBS series *Time Team America*. His doctoral work at Tulane University focuses on the Hollywood site, a Mississippian mound center located in Tunica County, Mississippi.

*Scott W. Hammerstedt* is a member of the research faculty at the Oklahoma Archeological Survey, University of Oklahoma. Current research focuses on the archaeology of late prehistoric Mississippian and Caddoan societies in what is now the southeastern and midwestern United States.

*Edward R. Henry* is a doctoral candidate at Washington University in Saint Louis. His research explores social, economic, and political institutions among small-scale societies. Specific interests include the entanglement of ritual/religious, political, and economic institutions, as well as transitions with complex hunter-gatherers and their increasing reliance on domesticated food production. His dissertation research investigates the temporality of geometric earthen enclosures in the Bluegrass Region of Kentucky with regard to general trends in the Ohio Valley. He uses geographic information systems (GIS), various remote sensing techniques (LiDAR, aerial photography, geophysical), geoarchaeology, and Bayesian modeling of chronometric dates to evaluate structural change of varying types and degrees across social landscapes in the Ohio River Valley. He coedited *Early and Middle Woodland Landscapes of the Southeast*.

*Jason T. Herrmann* is an anthropological archaeologist who uses remote sensing and field archaeology to study the interactions between ancient people and their changing environments, including transformations in landscapes, settlement systems, and land-use strategies. He is a research associate at Eberhard Karls Universität Tübingen as well as at the Center for American Archeology. Recent field projects include work in the Lower Illinois River Valley and southern Turkey.

*Jay K. Johnson* is professor emeritus of anthropology at the University of Mississippi, Oxford. He edited *Remote Sensing in Archaeology: An Explicitly North American Perspective* and *The Development of Southeastern Archaeology*.

*Jason L. King* is director of research at the Center for American Archeology. His focus is on the archaeology of the Lower Illinois River Valley, particularly mounds and monumentality, bioarchaeology, relatedness, mortuary practices, and ideology of Woodland period peoples. Since 2001 he has directed excavation and geophysical prospection projects at multiple mound sites in the Illinois Valley.

*Kenneth L. Kvamme* is a professor of anthropology and director of the Archeo-Imaging Lab at the University of Arkansas, Fayetteville. Most of his career has focused on archaeological prospecting, beginning with pioneering work in the 1980s that explored GIS-based models of archaeological distributions across landscapes. Since the 1990s he has pursued archaeological detection through geophysical prospecting and remote sensing, focusing primarily on the "big four" methods of magnetometry, electrical resistivity, ground-penetrating radar, and electromagnetic induction. Although his geophysical fieldwork takes him to projects worldwide, he specializes in the archaeology of the American Great Plains and Rocky Mountain West. He has authored more than 100 publications and 150 technical reports on archaeological GIS, spatial analysis and modeling, geophysical applications, and remote sensing. He is associate editor of the journal *Archaeological Prospection* and is on the advisory board of the *Journal of Archaeological Method and Theory*.

*Patrick C. Livingood* is an associate professor of anthropology at the University of Oklahoma. His research interests include the Mississippian period in the southeastern United States, including the archaeology of the ancestral Caddo in eastern Oklahoma and adjacent states. He is the author of *Mississippian Polity and Politics on the Gulf Coastal Plain: A View from the Pearl River, Mississippi* and coeditor of *Plaquemine Archaeology*.

*Jami J. Lockhart* is the coordinator of computer services, and director of archaeogeophysical and GIS applications for the Arkansas Archeological Survey. He also serves as research faculty in the University of Arkansas Department of Anthropology and as cooperating faculty in the Environmental Dynamics PhD Program. Lockhart's interests include archaeological research applications integrating digital technologies such as high-accuracy mapping, aerial and space-based remote sensing, multisensor geophysical surveys, GIS, relational database management systems, and exploratory data analysis.

*Sarah Lowry* is an archaeologist and geophysics specialist for New South Associates, Stone Mountain, Georgia. Her research interests include integrating near-surface geophysical and geospatial technologies for archaeological applications, geophysical applications in historical cemeteries, and Woodland and Mississippian sites in the Tennessee River Valley. She has published several articles on geophysical applications for archaeology and authored dozens of technical reports for federal, state, and private clients.

*Daniel P. Lynch* works in the cultural resource management industry and teaches anthropology part time. He is a doctoral candidate at the University of Massachusetts Amherst. His research interests include measuring the electromagnetic properties of archaeological soils and materials in both the frequency and time domain and the application of geophysical methods at historical archaeological sites.

*W. Stephen McBride* is director of interpretation and archaeology at Camp Nelson Civil War Heritage Park, Nicholasville, Kentucky. He has a special interest in the archaeology of both military life and African Americans. McBride has authored or coauthored numerous articles on the historical archaeology and history of Camp Nelson and other eighteenth- and nineteenth-century sites.

*Duncan P. McKinnon* is an assistant professor of anthropology at the University of Central Arkansas and a research associate at the Center for American Archeology. His research focus is on spatial dynamics and landscape archaeology of southeastern Indians. Current projects include a distributional analysis of Caddo art and ceramics and an examination of the constructed landscape of Woodland period occupations in the Lower Illinois River Valley. McKinnon is author and coauthor of several publications in *American Antiquity*, *Arkansas Archeologist*, *Caddo Archeology Journal*, *Midcontinental Journal of Archaeology*, and *Southeastern Archaeology*. He is editor for the *Caddo Archeology Journal* and was also a member of the PBS show *Time Team America*.

*Philip Mink II* is a research archaeologist and GIS coordinator at the University of Kentucky. He is an anthropologically trained environmental archaeologist whose research focuses on the role of cultural variation in human adaptability. He is interested in both landscape-level geoinformatics approaches to examining human and environmental interactions and in smaller-scale site-level considerations of place. Philip has been utilizing terrestrial remote sensing techniques for over a decade to answer anthropological archaeology questions in locales as varied as Grand Canyon National Park, southern Ohio, central Arizona, coastal Peru, Georgia, and Kentucky. At the Grand Canyon he has employed GIS modeling and terrestrial remote sensing surveys to investigate Pueblo period cultural variations and adaptations to this ecologically diverse natural wonder. In Kentucky he has conducted numerous geophysical surveys, sometimes as part of larger archaeological research projects, but just as often as a public service to the citizens of the Commonwealth.

*Tim Mulvihill* is a station archaeologist at the Arkansas Archeological Survey based at the University of Arkansas–Fort Smith. He has been involved with making new digital maps of multiple sites throughout Arkansas and has assisted with many geophysical projects across the state as well. Current research involves the late prehistoric and historic archaeology of the Arkansas River Valley in western Arkansas.

*Shawn M. Patch* is the principal investigator and director of geophysics at New South Associates, Inc. His research interests include archaeological geophysics, Mississippian communities in the Tennessee River Valley, lithic technology, and prehistoric and historic archaeology of the Caribbean. He has published several articles on geophysical applications in archaeology and authored dozens of technical reports for a wide range of state, federal, and private clients.

*Erin Pritchard* is a senior archaeological specialist with the Tennessee Valley Authority (TVA). Currently, her work focuses on planning and implementing TVA cultural resource programs related to archaeological resource management and protection pursuant to the National Historic Preservation Act and the Archaeological Resource Protection Act. She is the editor of *TVA Archaeology: Seventy Years of Prehistoric Site Research.*

*Amanda L. Regnier* is the director of the Oklahoma Archeological Survey at the University of Oklahoma. Her research interests include the ritual and social lives of the ancestral Caddo in the Red River Valley, the archaeology of Spiro and related mound sites in the Arkansas drainage, nineteenth-century Cherokee and Choctaw archaeology of eastern Oklahoma, and European Contact period archaeology of central Alabama. She is the author of *Reconstructing Tascalusa's Chiefdom: Pottery Styles and the Social Composition of Late Mississippian Communities along the Alabama River.*

*George Sabo III* is the director of the Arkansas Archeological Survey and a professor of anthropology at the University of Arkansas, where he also serves as codirector of the Environmental Dynamics PhD Program. His research centers on human/environment relationships, expressive culture (art and ritual) among southeastern Indians from pre-Contact to modern times, American Indian interactions with European explorers and colonists in the Southeast, and the anthropology of history in modern Caddo, Osage, and Quapaw communities in Oklahoma. His current projects include a study of protohistoric art, ritual, and social interaction in the central Arkansas River Valley.

*John R. Samuelsen* is the server administrator for the Arkansas Archeological Survey (AAS) and an anthropology doctoral student at the University of Arkansas. He began his work in computer applications in archaeology at the University of Florida. He then began working for the AAS creating websites, developing AMASDA Online, and researching Caddo archaeology using geophysical methods. His work at Spiro is part of a multistate effort to study past landscapes using geophysical methods along the Arkansas River.

*Daniel M. Seinfeld* is the Archaeology Collections and Conservation supervisor for the Florida Bureau of Archaeological Research, where he also oversees the state's unmarked graves program. His current research focuses on monumentality and ritual in the southeastern United States.

*Jennie O. Sturm*, a doctoral student in anthropology/archaeology at the University of New Mexico, is the owner of TAG Research by Sturm, Inc., a research and consulting company specializing in geophysical and geospatial applications for archaeology and the environmental sciences. She has conducted geophysical surveys and authored technical reports in 15 US states and five countries. In addition to consulting work, her research interests include using geophysics, remote sensing, and GIS analyses to study economic systems and prehistoric land use in the American Southwest.

*Victor D. Thompson* is an associate professor of anthropology and the director of the Center for Archaeological Sciences at the University of Georgia. His primary interests are the Native American societies and early European colonists who occupied the coastal and wetland areas of the American Southeast. The focus of his studies is to understand the long-term dynamics between humans and their environment and how these trajectories experienced ruptures and displacement at the moment of European contact. Most of his work has taken place in Florida and along the Georgia coast, although he has been involved in fieldwork in France, Mexico, Saint Kitts and Nevis, and Saint Vincent and the Grenadines. He is the author, coauthor, and coeditor of over 50 publications, including *The Archaeology and Historical Ecology of Small Scale Economies*. He is the founding editor for the book series *History and Ecology in Island and Coastal Societies* and the associate editor of book reviews for *American Antiquity*.

*Taylor H. Thornton* is an assistant director for numerous field schools at the Center for American Archeology and a doctoral student at the University of Toronto. Her interests are landscape archaeology, spatial analysis, geophysical prospection, and human-environmental interactions in prehistoric North America.

*Margaret Watters* is the northeast region (NER) National Park Service Archaeological Site Management Information System coordinator and works with the NER archaeology group on developing integrated approaches for helping national parks discover, record, preserve, and interpret their archaeological resources. She also consults with Visual Environment Solutions, LLC, in applied remote sensing for landscape exploration and modeling in archaeological, built environment, and media applications. Watters specializes in three-dimensional visualization of remotely sensed and excavated archaeological data for a new perspective on noninvasive modeling and analysis of archaeological sites. Watters has published and presented on applied research at sites around the world and has worked with numerous television productions. As a co–principal investigator and presenter for *Time Team America* on PBS, Watters supports archaeology as a gateway to science to engage and inspire today's youth in the application of science and technology and to help tomorrow's innovators and problem solvers define their future.

*Adam S. Wiewel* is a doctoral candidate in anthropology at the University of Arkansas, Fayetteville, where his research involves geophysical and archaeological investigations at Plains Village sites in North Dakota. He works as a research assistant at the Center for Advanced Spatial Technologies at the University of Arkansas, where he has undertaken numerous remote sensing projects, particularly those utilizing unmanned aerial vehicles, thermal imagery, historical aerial photographs, and digital photogrammetry. His research interests include remote sensing, archaeological geophysics, and Plains archaeology.

*James Zimmer-Dauphinee* is a researcher at the Field Museum of Natural History in Chicago. His recent research focuses on geophysical applications in archaeology, especially electrical resistivity tomography; however, his greater interests lie in archaeological applications of spatial analysis, remote sensing, mathematics, and computation.

# Index